HUMAN EVOLUTION
A SCIENTIFIC SOCIOLOGICAL
ANALYSIS

HENRY EDWARD MIDDLETON
B.A., B.Sc., C.Eng., M.I.Mech.E.

MERLIN BOOKS LTD.
Braunton   Devon

To
NORA

ISBN 0 86303 045-9
Printed in England by Phillips & Co., Crediton, Devon

CONTENTS

PREFACE

Because the subject matter of this small volume is very wide, some of the argument will appear unavoidably diffused; and also, in order to challenge some erroneous ways of thought which unfortunately have established themselves in social science, it has been necessary to use a very punctilious mode of discussion and analysis. This is perhaps unfortunate; however, the book is not meant for entertainment or light reading. The subject is effectively the establishment of a law of human evolution which is necessarily much more complex than that which applies to other species; therefore, the discussion is complex, and requires a good deal of concentration. One cannot remove the fatigue from this process. It may, nevertheless, kindle the interest to give some preview of what this book claims to achieve, namely:

The extension of the scientific principles of conservation of energy, uncertainty, entropy, systems analysis, etc. can be accomplished; though this volume is only a beginning.

Energy satisfies the necessary co-ordinating role for extending the methods of science to cover human activities. The 'free will v determinism' or 'action v systems' paradox is elucidated; and both rationality and morality are analysed and defined scientifically. These extensions of science, in my view, constitute verification of significant empirical referents.

These achievements do not require deterministic, machine-like, mechanical, optimisation which attempts to mould cultures to one stereotype. It requires plasticity of behaviour and of cultures, whereby men may fit into complementary niches with men, and cultures may interlock with cultures, in mutual regard and

5

appreciation. The concept of the human eco-system is relevant, in which conflict and waste are minimised and resources are proliferated by co-operation. Ideally conflict should be eliminated by sublimating it onto the higher intellectual plane as suggested by Novicow.

These are ideals for which men should aim; but, they cannot be attained to perfection because of man's dual nature in which, the lower animal biological mode intermingles with the higher rational mode of behaviour; and the higher strives to refine the resources for its fulfilment from the raw materials provided by the lower.

Conflict theory is, therefore, very relevant since it portrays the effects of the lower 'instinctive processes' whereby much of human life is 'played out'. Conflict also exists in every individual in whom the higher rational processes strive to master and refine the lower animal 'instinctive' raw energies to pursue ideal goals; nevertheless, it still remains true that ideals are the main distinguishing characteristic of man as a species.

The above is only a brief abstract, therefore it may also help to focus concentration if the reader quickly scans the summary (p92) before commencing the main body of the analysis.

1

INTRODUCTION

The purpose of this short volume is to outline the overall plan for a scientific social theory of human activity and evolution. This is based, centrally, upon the principle of conservation of energy (C.O.E.), but also utilises other complementary scientific concepts such as indeterminacy, entropy, dynamic self-maintenance and systems analysis.

This book does not pretend to deal exhaustively with all of the problems involved: it merely attempts to identify them and set them into a scientific context. There are three reasons for the brevity: firstly, people have not the time to read mammoth books: secondly, if a theory is to be widely understood it must be expressed in a maneageable space and form: and thirdly it is hoped that this volume will provoke discussion, criticism and further thought which will benefit future work.

When dealing with social science, questions of philosophy will inevitably arise; nevertheless philosophic discussion should not be allowed to escalate for its own sake. There are more than enough philosophies in existence to ensure that they can be selected and mixed as the writer thinks fit. Whatever the philosophy chosen it must, in some way, underpin the facts as he sees them.

Therefore, there should be an interdependence between a writer's philosophy and his science; it is better that this relationship should emerge rather than be prejudged. In social studies some of the most fundamental facts, such as man's ability to choose and act morally, are both matters of fact and also of philosophy. Opinions differ or the facts will appear differently to different theorists according to their philosophies or beliefs.

Every theorist brings into his work, at the outset, his own prejudices or beliefs. These should, in the longer run, be examined ruthlessly in the light of his final conclusions. Readers are entitled to infer the existence of such influences and to make reasoned criticisms of them.

However, philosophy should be related firmly to the relevant social facts. One should, ideally, start with the facts and attempting the widest possible generalisation from them, arrive at a provisional integration between his philosophy and his science: then, ideally this should be critically examined and refined until by repeated re-examination one is finally convinced that here is a set of unifying ideas or principles which can form a basis for an overall theory which accounts for all of the social facts.

In practice, prejudices and beliefs are often inculcated in the form of doctrines, dogmas, creeds or cultural habits long before the critical faculties are formed. Such prejudgements need to be identified in one's mind, and very carefully and critically examined. Nevertheless, since it is in the light of facts that philosophies must be judged, one cannot finally appraise philosophies until the facts and their related theories have been critically examined and expounded.

From the above, it can be inferred that there should be a constant interplay between one's social theory and philosophy. There is nothing to be gained by adhering to a philosophy that is at odds with the known facts; for example, there is no point whatever in persisting that scientific laws must act deterministically upon man when it is quite clear that men can and do make choices. Some theorists insist, against the weight of all the evidence, that science is invariably identifiable with determinism: on the contrary, the principle of indeterminacy is relevant to social science and also to some other areas of science. Science is a search for truth and much of the truth cannot adequately be explained under a philosophy of determinism or a theory of automatic response.

It is not my policy to become involved deeply, in this book, in the philosophy of ultimate realities of matter and energy; the further one strays from the social life of man the less one is able to maintain a clearly relevant scientific model. However I shall not attempt to conceal my belief in Catholic Christian philosophies; they appear to be perfectly compatible with the scientific social facts. One must not, of course, derive science from philosophy, although philosophies which are sound can be verified by study of the social facts. Readers will judge for themselves which course I have taken.

One could not produce a serious and responsible social theory without having in mind an equally responsible philosophy. I hope to discuss philosophy in a later volume. For the moment it must suffice to say that philosophy must be always responsible, not carried on in a vacuum, but interwoven with an adequate scientific social theory. The existing disastrous dichotomies in man's thought and actions must be bridged before he can advance beyond the status of a very cunning animal, and so progress further along his higher rational mode of development. This is a direct challenge to those who believe solely in conflict theory.

2

THE RELATIONSHIP BETWEEN SCIENCE AND SOCIAL STUDIES

The title of this section has been chosen so as to avoid approaching the subject in the form of a question which might, in some way, prejudge or narrow the choice of answer. The type of question which should be avoided is that which asks whether sociology is a science; the danger lies in the temptation to answer either yes or no to this question when it cannot be answered in such a clear-cut way. In fact, as was mentioned in the foregoing introduction, there can be no sharp division between science and philosophy in the social field; but rather there is an intimate interdependence between them since social phenomena can partake of both.

Those who insist upon a sharp division between social science and philosophy must ultimately rest their case upon the faulty assumption that science deals only with matters in which deterministic laws apply exclusively; in this case, it would be possible to identify science by its ability to enable prediction of events predetermined by these laws. Phenomena which cannot be predicted would then fall automatically in some other discipline.

The error in this assumption lies in the fact that in some areas of science the principle of indeterminacy or uncertainty is applicable. As an accepted scientific principle this is essential to quantum mechanics; but indeterminacy is also applicable to biology in even more obvious ways. There are profound philosophic discussions today in microbiology as well as in quantum mechanics, but in the affairs of men indeterminacy attains its highest level; it is quite clear that the philosophic problems of indeterminacy are vitally involved at both ends of the evolutionary scale. Therefore, one cannot separate philosophy and science at the fundamental level where these two disciplines are used to elucidate the problems of man in society; and the principle of indeterminacy applied to man's behaviour in no way departs from the methods of science.

The principle of indeterminacy clearly raises the vital issues of how men should choose to act and upon what standard of thought and rationality. These issues again encroach upon morals and philosophy. These cannot be dealt with exhaustively in this small book; nevertheless they are related to the scientific principles and facts or truths. Therefore it appears that the question of whether sociology is a science is no more than a play upon verbal definitions, since it all depends upon one's definition of science. If one adopts a narrow definition in which only deterministic phenomena amenable to precise prediction are accepted as science, then much that is usually included as science would have to be consigned to other disciplines: however if one adopts a wider definition and includes all observable aspects of fact or truth, then science includes not only things that are determined and predictable but also things that are uncertain and unpredictable. This view of science covers not only laws of general application like conservation of energy but also the principle of uncertainty. In social science it is the interdependence and complementary nature of these apparently opposed principles that is important.

In summary the wider view of science deals with facts and truth in every observable aspect: but of course when we begin to associate science with the more general search for truth, then it merges with philosophy and morals.

However the scientific element in the marriage between philosophy and science is vitally important. It must be appreciated that where the principle of uncertainty applies precise prediction of events as empirical referents cannot be used. In these cases it is the ability to extend the use of principles of the more precise sciences into the field of the less exact that serves as verification of scientific validity. It is the rationality involved in this process of extension, its adherence to established scientific principles, that is essential. Rationality is essential to science but deterministic reactions are only sometimes applicable.

In the process of rationally extending scientific principles into the uncertain field of human activity, it is the scientific principles themselves that must form the basis of the rational procedures; and in particular the law of C.O.E. must provide the unifying role in logical deductions. This is of itself the only empirical test of validity, namely that the logic shall be consistent with established scientific principles.

Therefore, the essential function of science is to set up a model that represents effectively as adequately as possible the phenomena observed. If the facts under observation proceed according to deterministic laws then a deterministic model is appropriate; however, if there are elements of uncertainty these must be represented by corresponding indeterminacy in the model. In the former case precise prediction for empirical referents is possible; but in the latter case we must rely upon rational deduction based on accepted scientific principles. In either case it is the appropriateness of the model to the representation of the facts that is the essential mark of science rather than determinism which is not a universal attribute of science.

A new theory is acceptable as a replacement for the older one if it accounts for a larger range of the phenomena than did the older theory. This kind of development can be achieved, either by discovering new laws and relationships or by logically extending the use of established scientific principles to new fields.

The latter is the method used in this analysis; it is, of necessity, in social science, a reflexive process; but it is claimed that it extends the use of scientific principles and also throws new light upon three vital areas of social activity; namely freewill, rationality, and morality.

This concludes a brief preview of some of the fundamental ideas that are used in this analysis; they will not find favour with every reader. I have not worked through full explanations of the scientific concepts used in order that the overall picture will be seen unobscured by the detail of supporting evidence. Therefore, it is now necessary to look in more detail into some important aspects of science.

3

THE METHOD OF SCIENCE

There is no generally agreed definition of science; there is lively controversy as to whether social and historical study can be treated scientifically. In this connection it is instructive to consider the wide variety of disciplines and intellectual activities which have been accepted as scientific; these have included mathematics, logic, astronomy, physics, chemistry, geology, biology, zoology, botany, etc. It is also instructive to examine the various criteria that have been used to classify the sciences. A common distinction is made between exact and descriptive sciences; physics is an example of the former and zoology of the latter. This contrast shows that exactness is not, of itself, an essential of science; in fact exactness is generally associated with the possibility of accurate measurement. However, measurement operations are always subject to some uncertainty or error; this is most especially so in quantum mechanics in which the principle of uncertainty is vitally important.

Repetition is a very potent method of checking the correctness of a scientific theory; but it is only in the more exact sciences that it can be achieved. Repetition is not, therefore, essential to scientific method.

Experiment is also one of the most powerful tools of many sciences. The experimenter artificially varies the conditions under which phenomena occur; this may greatly increase the frequency of certain rare but important situations; and thus helps understanding and analysis. Experiment is sometimes used in social science; but there are observational sciences such as astronomy and geology in which the setting up or repetition of situations at will is impossible. Therefore, experiment is not essential to scientific method.

Some method of classification of the phenomena to be described is necessary in order to avoid the enormous labour of treating each individual example as a separate class. This economy of description, as for example in the classification of plants into species, is perhaps one of the most generally necessary methods of science. The method of classification is much more useful if some theoretical understanding of the matter subject to classification is attained. Therefore theoretical analysis and speculation confirms the effectiveness of the classification used; and also of rationality and understanding which is the most universal and essential attribute of science.

It is valuable, at this point, to examine the nature of scientific rationality: this involves methods of concept formation and inference and techniques of controlled observation. The logical part of scientific method consists of rules of vocabulary and sentence construction as well as rules for inferring conclusions from data; scientific language is more precise than ordinary language since its statements can usually be verified; new terms and relationships may be created by defining them in terms of scientific concepts already in use; it is, therefore, a fundamental characteristic of science that it builds its language, concepts and theories upon the basis of those scientific concepts that have been already established. Wherever possible concepts are defined in terms of precise observation or measurement; and personal or emotive meanings are eliminated.

In the most highly developed phases of science many specific theories can be deduced from a few basic principles. Rationalist philosophers and scientists consider that deductive inference is more conclusive than inductive inference since the conclusion of a valid deductive inference follows logically from its premisses whereas that of an inductive inference does not; this underlines the role of rationality as the most fundamental attribute of the scientific method.

Finally let us examine the claims of social studies to rank as scientific; in the past, they have not built very directly upon the existing concepts of science; and again it is difficult to claim much by way of precise measurement, controlled experiment or even of any agreed system of classification. It is the contention of this book that there can be no virtue in any false claims to preciseness of prediction or automatic determinism in man's affairs; and there are great complexities of classification also involved. However, from the above discussion it appears that the principal attribute of science is rationality using deductive inference from premises based upon existing accepted scientific principles and concepts. There has been no serious concerted attempt in this direction; it is the specific purpose of this book to outline an overall plan to remedy this deficiency. Towards this end some of the most relevant scientific principles will now be discussed briefly.

4

CONSERVATION OF ENERGY

The principle of C.O.E. states that energy can neither be created nor destroyed but it may be transformed from one form to another; thus the total energy in any system will always remain constant provided it is a closed system which allows no gain or loss across its boundaries. There are many everyday examples of energy transformations between the forms of mechanical, electrical, heat and chemical energy; it was also established by Einstein's theory of relativity that the apparently solid matter of the universe is built up from an energy basis, so that there is an equivalent of energy to every mass of matter. Energy, therefore, is an unavoidable influence and constraint upon every human material activity; and hence, men live in an environment composed of energy with bodies also of energy. Energy is the material medium within which they must live and act.

It follows from the limitations of the energies in any closed system, that the energies or resources of man in any situation must be similarly limited; from this fundamental limitation in the very origins of matter and energy, arises also the equally fundamental scarcity of man's economic resources. All matter is built up from energy; and similarly all the commodities of men are constructed by the energies or labour of men upon the energies of matter. This is more than an analogy between the concept of relativity and man's productive processes, since it identifies the fundamental common basis in both.

This fundamental factor of energy was recognised by Karl Marx in his labour theory of value which states that; the value of any commodity is determined by the quantity of average socially necessary labour required to produce it. I would only add that the energy in question need not be solely of human bodily source but may also be from other sources of power harnessed by machines to man's control. Nevertheless, the concept of energy is by no means precise or deterministic in its application: all theories of value have in them some degree of uncertainty as to how value relates to monetary considerations and price: every arbitrary payment of a price is someone's assessment of a value; but it must be admitted that they rarely involve any very precise energy calculations. However, one does not expect to find precise predictions in human decisions; it is the consequences that arise from the energy constraints, which manifest themselves in scarcity of resources, that are significant irrespective of whether they are predictable. Monetary value is, therefore, only a very rough symbolic representation of the inherent energy aspect of man's activity.

In general, man's thought proceeds by symbolic representation and this has usually evolved rather than been precisely devised. Indeed, it is the precise logical devising of symbolic representation that is the one indispensible aspect of science. In social science the logic behind symbolism must be made as precise as possible. That is why the use of precise concepts, already established in the 'exact sciences', is now essential for the further advancement of social science. This necessary preciseness is required in the scientific logic that must be applied to devising the symbolic model and not to any falsely assumed deterministic or automaton-like idealisation of man's behaviour.

Thus the principle of uncertainty is complementary to that of C.O.E. in social science, as also is the case in some branches of the more 'exact sciences': this now requires some further brief discussion.

5

THE PRINCIPLE OF UNCERTAINTY

This principle amounts to the admission that in some areas of science we are unable to obtain precise knowledge of the phenomena under observation.

In the field of quantum mechanics, the following statements from *Man and Science* by Professor W. Heitler illustrate the principle of uncertainty: (first edn 1963)

'In quantum mechanics statements about probability take the place of the definite deterministic predictions of classical mechanics.' (p.38)

'In general there exists only probabilities for the expected results of measurement and each such measurement influences the object.' (p.39)

'A position measurement means a very drastic interference with the atom, which in some cases can even lead to ionisation, i.e. the destruction of the atom as such.' (p.40)

'To ionise the atom a considerable expenditure of energy is necessary. Since the position measurement may, as we have seen, ionise the atom in some cases, it is clear that a position measurement changes the energy of the atom . . . . the energy is now no longer sharply defined and does not have a definite value.' (pp.40-41)

'We have here arrived at the most important basic principle of quantum mechanics. We can measure a quantity which was uncertain, and then it becomes, by definition, sharply defined. But the knowledge thus acquired is at the expense of the definiteness of some other quantity. Something is always uncertain . . . . In an intact atom the energy is sharply defined, everything else is uncertain. We measure the position: this becomes sharply defined, while the energy becomes uncertain . . . . We call two quantities which cannot be simultaneously sharply defined, complementary . . . . The indeterminacy of quantum mechanics is the direct outcome of the principle of complementarity.' (pp.42-43)

'The single atom is thus an object of very abstract character that does not admit of a complete description in space and time.' (p.45)

This can be summarised thus: One cannot adhere to a science of determinism in quantum mechanics; nevertheless, scientific methods and in particular the principle of C.O.E. are very effective tools in analysis of quantum mechanics. At the base of this principle of uncertainty or indeterminacy in quantum mechanics lies the impossibility of precise measurement.

In a similar, but much more complex manner the principle of uncertainty also applies to the activities of man; in social science the diffculties of measurement are multiplied but the overall constraint of C.O.E. nevertheless, still operates. It is, therefore, vitally important to utilise this most general of scientific principles in order to establish analytical order and system in social science.

In attempting to make measurements and observations of social activities man is also involved in such a way as to change the object under observation. This is again similar to the involvement of man in the measurements of quantum mechanics but it is much more subtle and complex; man does not just unwillingly modify the system, as in quantum mechanics; he is in fact inevitably involved as the very basis of the system under observation.

However, it must be emphasised that there is no suggestion intended, in this book, that man's life processes are at a level similar to that of the atom. All that is intended here, is to apply scientific principles of great generality to the analysis of social science. Thus far the general principles of C.O.E. and uncertainty have been discussed: it is now necessary to mention a third very general scientific analytical technique, namely systems analysis.

6

SYSTEMS ANALYSIS

It is difficult to find one, generally accepted definition of systems analysis; it is, in my view, the modern scientific way of thinking in an organised systematic way. Systems can be applied to predetermined or deterministic phenomena but it is the advance of science into the areas of the indeterminate, uncertain and unpredictable that has made it essential to develop the new and more flexible way of thinking which goes under the title of systems analysis. The older deterministic way of thought was adequate for dealing with the sciences that are precise, determined and predictable. The newer mode of thinking can also deal with such relatively simple problems; but, because it is wider in scope and in generality, systems analysis can also be used for those problems that are uncertain, indeterminate, and unpredictable.

In the uncertain areas of the newer sciences, such as social science, stability of a different order must be considered; the older concept of equilibrium is replaced by a newer term, namely, dynamic-self-maintenance which can take two possible forms; these are defined as homeostasis or morphogenesis.

Homeostasis

Cannon defines homeostasis as 'the ensemble of organic regulations which act to maintain the steady states of the organism and are effectuated by regulating mechanisms in such a way that they do not occur necessarily in the same, and often in opposite, direction to what a corresponding external change would cause according to physical laws. The simplest example is homeothermy.' (p.35 *Systems Behaviour* edited by Beishon and Peters) Homeothermy is the feedback process whereby temperatures in animals is maintained constant.

Morphogenesis

In his article, 'Society As a Complex Adaptive System', Walter Buckley observes 'In dealing with the socio-cultural system, however, we need yet a new concept to express not only the structure-maintaining feature, but also the structure-elaborating and changing feature of the inherently unstable system. The notion of "steady state", now often used, approaches the meaning we seek if it is understood that the "state" that tends to remain "steady" is not to be identified with the particular structure of the system. That is, . . . in order to maintain a steady state the system may change its particular structure. For this reason, the term morphogenesis is more descriptive.' (p.158 *Systems Behaviour*, Beishon and Peters)

To summarise; the significance of the concepts of homeostasis and morphogenesis lies in that they indicate ways whereby social groups or organisations can survive in situations in which actions are not automatic or determined, but can be calculated towards the conscious purpose of survival or even growth of the group. This purposeful effort for survival requires the maintenance of systems of organisation and control; − in this connection, the concept of entropy is relevant and will now be briefly discussed.

7

ENTROPY

Entropy is one of the most important concepts of systems analysis; it is applicable to all physical systems. Any closed system tends to move towards a chaotic or random state as the potential for energy transformation or work decreases towards zero. This disorder or randomness is known as the entropy of the system. This concept of entropy is vital in the distinction between closed and open systems; closed systems tend towards increasing disorder or randomness over time which means increasing entropy.

However, biological and social systems are not closed systems; they have various inputs and outputs; the inputs are in the form of varieties of material energy which are transformed and so increase the potential so as to offset the process of increasing entropy and randomness.

Closed systems are not living nor are they social systems. If an open system somehow loses its supply of material energy from outside, it can no longer maintain the potential of its internal parts; and consequently it will also lose its dynamic equilibrium with its surrounding environment; and eventually it will fail to survive. It is this relationship between maintaining energy potential, prevention of disorder, and prevention of entropy increase, that is vitally important for the survival of any open system and its maintenance in a state of dynamic equilibrium.

Negative Entropy

In a closed system the change in entropy must always be positive since both entropy and disorder continually increase until finally the system stops. However, in the open biological or social systems, entropy can be arrested and may even be transformed into negative entropy, a process of change to a state of more complete organisation. The only way in which the open system can offset the tendency towards increasing entropy is by continually importing material, energy and information in various forms, transforming them and redistributing unwanted resources to the surrounding environment. These processes of dynamic self-maintenance can only be achieved by the use of adequate controlling systems which can vary from the simplest organisms to the highly complex human brain or the intricate social controls.

Feedback Mechanisms

The concept of feedback is essential to the understanding dynamic equilibrium in systems. Feedback continually provides information from the environment which enables the system to adapt and survive.

Negative Feedback

This provides data which indicates that the system is deviating from required directions and therefore, needs to readjust its internal processes to a new steady state. These negative feedback processes are an essential factor in the achievement of survival by homeostasis or morphogenesis. The reactions in organisms whereby they maintain constant temperature under changing environmental temperature is one example of this.

Equifinality

In systems analysis this relates to open systems. The principle of equifinality states that final results may be achieved with different initial conditions and in different ways. This is illustrated by the geographical principle of possibilism which recognises that a variety of social cultures can exist in similar geographical situations; thus the social system is not completely restrained by the simple cause to effect relationship of closed systems which are subject to increasing entropy until the entire system stops.

The above is a very brief account of those scientific principles that are most important and relevant to social science. In general they are not deterministic laws. I hope that this will suffice to deter, in some small measure those who misguidedly seek for deterministic activity in human affairs and also those who erroneously identify science with deterministic reactions only.

THE HYPOTHESIS STATED IN GENERAL TERMS

Man is the latest stage in the long process of evolution which is assumed to have commenced with completely random movements of inanimate matter. The subject is controversial, but it is the postulate, in this thesis, that evolution is progressing towards less random, more organised, more purposefully goal-seeking and increasingly rational activity, with man as its latest development.

If these assumptions are correct, it follows from the scientific principles previously stated, that man in society, is an open system; and therefore, social systems survive or grow by obtaining a supply of material energy from their environment sufficient to maintain or increase the potential of their internal parts. In this process of dynamic self-maintenance, entropy and randomness are either decreased, or are, at least, prevented from increasing.

However, it would be naïve to suppose that such progress towards increasingly rational activity were taking place continuously. One must always beware the oversimplified explanation of this kind, since it neglects the capacity of man to choose either what is rational or what is irrational; and it neglects also, the fact that man may exert decisions of choice either with or without full knowledge of what is rational.

Indeed, the above postulate of increasing rationality in man can only be seriously considered as an overall, general trend rather than as a deterministic or automatic movement. Since man has far from attained complete rationality, it follows that, at any moment, his actions can be both partly rational and partly irrational; rationality is the ideal which is never fully achieved. Therefore, the postulate must be elaborated in order to produce a model that can represent both of these existing aspects of man's behaviour.

In my view, it is the attempts to produce over simplified social models couched in over-simplified deterministic language forms, that falsify and ruin social theory. Such attempts distort the facts of human life, in that they neglect to make due allowance for the uncertainties that surround the decisions of men; they neglect the scientific principle of uncertainty at the very point where it is most relevant, that is in the activity of man, the most variable being in all creation.

There can be no conclusive proof that an initial completely random state of matter ever existed; but it is an assumption that is widely held as a reasonable scientific postulate. Neither can there be any guarantee of what will be the next stage in human evolution. We do not know whether man will become increasingly rational in his social life or whether, perhaps, he might disintegrate his own social structure in a nuclear holocaust. All one can do is to speculate upon what might appear to be the most rational, and therefore, the most desirable course for man's future development.

The vital point to be emphasised here is that man can affect, by his decisions and actions, the shape of his own future; he is not only an observer of social life, but also a participator in the action who is able to change the existing social situation into something different in the future. Therefore, there arises a serious moral problem of what one feels should be done and what principles should be the motive for human activity. The answer to this is quite clearly that man should strive to become increasingly rational in his activities; nor is it satisfactory to leave the term rationality undefined: unfortunately, it has sometimes been dismissed tritely, in three words, as 'means to ends'. In order to avoid this kind of futile oversimplification, a later chapter of this book defines and discusses rationality at much greater length. However, as this present chapter considers the hypothesis in general terms, a few comments in the same general vein will also be made about rationality.

Having made certain provisional reservations and cautionary remarks, the evolutionary concept can be modified in the following general way:

Man is now in a stage of evolution in which his activities and motivations have much in common with the rest of the animal species. This is hardly surprising, since he is assumed to have evolved from other animal species: and man is, indeed, himself an animal.

However, man is uniquely different from every other animal in the scope of his rational capacity. Of course it must be admitted that some other animals have a certain, relatively small rational capacity; but this is very slight indeed when judged upon the results!

Which other animal has produced any result that compares with Einstein's theory of relativity, Shakespeare's plays, the poetry of Wordsworth or the art of Michelangelo? There is in these human achievements a complete difference in kind and an exploration of dimensions of reality unknown to other animals: and the intellect of man is so vastly different from that of other animals that one is justified in the belief that it possesses an entirely higher quality. For this higher quality in human thought and feeling there is no better word than spiritual; it is higher and more profound than the potential of any other animal: it is clearly a higher dimension.

If it were not so, why then should the other animals not be able to develop it also?

Before my critics point, with vehemence, to the atrocious cruelty and stupidity of much human behaviour, it is worth a reminder that I have quite carefully conceded the fact that man is still an animal with motivations that are frequently no higher than those of other animals. How might they perform if they possessed the intellect of man? Would they always choose to act rationally, or might they not often behave at their lower rather than the higher level?

Those who are determined to believe in deterministic human reactions will never be moved by any logical argument; but in my view they are setting their faces against the clear facts.

In addition to being an animal, man also has the capacity to aim towards an evolutionary goal of increasing rationality; although he may neglect the opportunity; and indeed, perfection in rational behaviour is without doubt unattainable for animal species. This defines the human problem in a nut-shell. He is both animal and higher rational being; which way will he choose to take? This is the essence of what must be incorporated into our scientific model of man in society.

Thus the lives of men contain two intermingled modes of evolution; firstly, that of the animal kingdom from which he is derived; and secondly the higher mode of increasingly rational behaviour which is distinctively human and beyond the capacity of other animals.

The distinction suggested above corresponds generally with that which used to be made between instinctive and intelligent behaviour. It was generally concluded that these two modes must be interdependent in man; and that is exactly the way their relationship is intended in this thesis. In fact, the instinctive or lower animal mode of behaviour is considered to be the source of the raw material from which the higher rational mode must be created. This is clearly in accord with the concept of crude energy being acquired and transformed into a more highly organised and refined form by the human organism and social system. Or to express the point in a slightly different form the lower animal nature must be controlled to serve the higher rational mode of life.

It will be observed that in this context I have identified the higher rational mode with a higher spiritual dimension. This will be discussed further in the later chapter on rationality; for the moment suffice it to say that whatever is good must also be rational, if the whole truth is known.

The increasing development and extension of rationality requires the discovery of universal laws or principles that can be used to relate conceptually, all multifarious phenomena of man's environment. Some such laws have already been discovered in the form of laws of C.O.E.

Scientifically these laws have withstood very rigorous tests and have been in constant scientific use; their extended application into the field of social theory must itself constitute significant progress in man's rational mode of action.

In the process of rationally extending scientific principles into the uncertain field of human activity, it is the scientific principles themselves that must form the basis of the rational procedures; and in particular the law of C.O.E. must provide the unifying role in logical deductions. This is of itself, the only empirical test of validity, namely that the logic shall be consistent with established scientific principles. The main principles used in this way are those of C.O.E., indeterminacy, systems analysis, and entropy.

The higher rational human mode of activity is distinguished by the continuous acquisition of resources from the environment; and constant striving for greater efficiency in the use of scarce resources and energy; these processes have the effect of promoting dynamic self-maintenance in the human systems by conserving resources, promoting organisation and preventing rundown by disorder and entropy increase.

In the process of rationally conserving energy and resources, when activity becomes less random and more rational, the number of trials to achieve a required result is reduced and also the consequences of men's actions become more predictable; this makes for economy of effort, which again underlines the significance of the concept of C.O.E. in human affairs.

Man's higher rational activity is intimately bound up with symbolic representation and abstract thought processes. In these processes the lower animal form of evolution based on 'survival of the fittest' and elimination of the weaker animals, gives way to a situation in which it is the ideas which must prove their fitness to survive; those less effective in helping man to acquire and conserve scarce resources of energy and material, for rational purposes must be relinquished. Thus it can be seen that man's higher rational mode of activity involves the use of symbolic models, plans, concepts, theories, and the setting of objectives or ideals; these also involve some freedom to discriminate and choose actions consistent with conservation of scarce energy and resources.

It is, of course, quite fair to query whether any of this potential higher rational mode of activity ever produces the fruits that it might appear to promise; for is it not evident that it is merely perverted, largely, into increasingly cunning and destructive forms of conflict and wastage of energy and resources? It should, by definition, be based upon conserving man's energies; but his higher rationality is, all too frequently, employed in the wastages of conflict rather than devoted to the economies of co-operation. This intimate interweaving of the two modes of human behaviour is of prime importance in the construction of a scientific model of human evolution. Nevertheless, one must never forget that it is the ideals and moral qualities of man that are his truly distinctive characteristics; since these set him apart from all the other animals.

Therefore, continuing in the same general vein, a few more general remarks will illustrate the nature of the battle between man's two different modes of action, in recent history.

In the last hundred years, and particularly with the 'welfare state', the efforts to preserve human life and dignity have markedly increased. Unemployment and starvation are no longer regarded as humanely acceptable sanctions in the competitively economic context: they are not now regarded as the automatic regulators of human society as was suggested by Malthus. However the elimination of such harsh sanctions has led recently to an urgent search for new ways, theories and concepts for the regulation and planning of social and economic affairs, a search, indeed, for new and more effective ideas.

All this is in accord with the theories of Novicow which suggest that human competition should become less physiological and more intellectual as time proceeds. Admittedly the ideas of Novicow are very optimistic and idealistic; however, such setting of ideal goals is the very essence of what is described in this analysis as the higher rational mode of man's development; and who will deny that this is one of man's most distinctive characteristics?

Therefore, it may be argued that increasing rationality tends towards increase in size and complexity of group co-operation; and within such groups a decrease in physical conflict might occur, although intellectual competition would perhaps increase. It must be emphasised, that these desirable effects would only occur if rationality were to continuously increase.

It is also arguable that as the groups grow larger, so the conflict between them also increases in scale and in destructive capacity. There is a great deal of evidence that both of these tendencies are present in man's social life; and this supports the view, expounded in this book, that there exist two differing modes of human evolution which are interwoven into man's social activity; the higher being based upon rationality, co-operation and conservation of energy; while the lower is founded upon animal tendencies towards conflict and struggle for survival which causes wastage of man's scarce resources.

In general, it appears logical to identify co-operation with the more rational mode of human behaviour; since it can increase economy of effort by efficient division of labour and also reduce the waste due to conflict.

This chapter has given some very brief general comments upon the hypothesis of human evolution based on scientific principles, and especially upon the law of C.O.E. It leaves a good deal of detail to be discussed in later chapters and indeed in later volumes.

A very serious attempt has been made, in the preceding pages, to avoid the pitfall of determinism; however, it would be naïve to claim that this attempt had been completely successful. Most of the discussion was, of necessity, of energy and the more material aspects of things, since these are the aspects which lend themselves to scientific treatment. Material gain has been seen as resulting from co-operation and elimination of wasteful conflict. From such arguments it might easily be inferred that material gain is the primary purpose of man's activity. Such an inference would, in my view, give a falsely materialistic and almost deterministic bias to the overall analysis. By that I mean that if man is so biased towards material gain then his freedom of choice is something of an illusion.

The whole emphasis should come from quite the opposite direction, co-operation should be primarily motivated by feelings of mutual regard, legitimate desire for social acceptance, and the instincts for caring and belonging. Once these become the basis for individual and social attitudes, then co-operation and material gain will follow as a logical consequence.

It is to be hoped that this objection to material gain as the prime mover in human activity will not be regarded as unduly subtle or philosophic; since co-operation for the sake of selfish individual gain is always liable to be easily corrupted into conflict which is quite the opposite of what is desired. This will be discussed in greater detail in a later chapter but I mention the matter briefly here in order to avoid being inadvertently carried into the insidious pit of determinism.

In the same context, it is also worthy of mention that as well as some finer feelings of mutual regard man can have a regard and feeling for the truth concerning himself and his environment, for its own sake: he is by no means tied inevitably to the pursuit of material gain for selfish reasons.

9

FURTHER DISCUSSION OF MAN'S TWO MODES OF EVOLUTION

There is no doubt that man is an animal, but in addition he also possesses certain outstandingly rational qualities which distinguish him from other animals. It is important, therefore, to explain this higher more rational mode in relation to the lower animal nature of man. Towards this end, the historical processes of development in Britain during the last two hundred years are useful as illustrations.

The theory of Malthus serves as an effective datum for the beginning of this period and also a foundation for the concept of man's animal mode of behaviour. The Revd Thomas Malthus wrote his theory on population in 1798; it painted a somewhat dismal picture depicting a conceptually idealised version of how man's population might have been controlled in a similar way to that of other animals; that is to say, population being controlled without anything akin to rational interference with the basic mating instincts as in animals.

It is extremely significant that this most acutely analytical work inspired not only Darwin's Theory of Evolution but also the nineteenth century utilitarian philosophers, economists, as well as the poor law and social policy for the whole of the nineteenth century. Malthus was often misunderstood and this occurred because his scientific approach to model building was far too advanced for his contemporaries.

Carr-Saunders points out in his book *Population* pp.23-24: 'Malthus . . . used certain phraseology upon which undue attention has been concentrated. Population, he said, would if unchecked increase in a geometrical ratio, whereas the means of subsistence . . . could not possibly be made to increase faster than in arithmetical ratio. Opinions differ as to the importance which Malthus attributed to the ratios themselves; some think that he attached great importance to them, while others hold that he used them rather as an analogy than as an exact statement of the facts.'

In my view, Malthus attempted, and largely succeeded in setting up a model of human evolution involving two interwoven modes of activity, one being the lower animal, and the other being the higher rational mode. He clearly predicted the kind of disastrous consequences for population and living conditions which would occur if men did not devote rational thought and self-control to the problems of population. Of course, Malthus was unable to predict all of the trends in population that have occurred since his time; and in view of the indeterminacy in human affairs, one should not expect such predictions. The important fact is that he constructed a model with all the right ingredients to represent man with all his animal imperfections and also his higher rational aspirations and potentialities: many subsequent philosophers, theorists, and social reformers have been inspired by the wonderful insight of Malthus's work. It may be that he fell partly into the trap of determinism but it is a very difficult trap to avoid since the very construction of ordinary language encourages the simple direct causal relationships. Nevertheless, Malthus did strongly assert that man should exert his rational choice and by self-control solve his population problems.

In general, Malthus produced a model of humanity which was realistic and true to life. My own model is almost identical with his; the difference lies in degree of awareness of the model-making process. Malthus could not, at that time, have thought of himself as model-making, since this scientific concept did not exist in his day; nevertheless his grasp of logic was so sure that he was able to identify the essential ingredients for a very representative model of human evolution.

In this book I am simply stating quite explicitly the essential ingredients of the model which Malthus left, to some extent implicit. In any case Malthus would not have felt too happy with the use of the ideal types such as lower animal and higher rational. This, in my view, is a normal and legitimate aspect of scientific model-making which was not developed in Malthus's day. The fact that one cannot expect to find such ideal types in real life does not detract from their usefulness in the detailed scientific analysis. No one expects to find a man who is purely animal nor yet one who is perfectly rational.

Ideal types can be used as parts of an overall model provided we realise that in practice the ideal types will not exist separately but will, if present at all, be intimately interwoven into the whole system represented. Later in this chapter this intimate relationship will be illustrated by a description of the construction of the human brain; this will provide physical evidence for the existence of different portions of man's brain where the two different modes of thought and action could be generated. Observation confirms that men can act, apparently upon impulsive instinct or alternatively with sublime rationality and self-control. It is because these parts of man's brain are distinct but interdependent that they can originate the two different but intimately interrelated modes of thought and action.

The above quotation from Carr-Saunders indicates that Malthus postulated a category of human behaviour closely approximating to an 'ideal' type of animal activity in which there would be no thoughtful control of population. He also discussed what needed to be done by way of intelligent control of population; thus he recognised a higher, more rational and distinctively human mode of behaviour. Admittedly Malthus did not argue the case with complete awareness of the analytical techniques which he was employing: nevertheless, his language and logic were quite coherent within the context of his knowledge at that time; and his conclusions have been confirmed by later more precise analysis.

As Mr Harold Wright has said, (p.24 Carr-Saunders *Population*):
'Unfortunately for the human race, the essential validity of the Malthusian principle of population is not destroyed by the substitution of an accurate account of the growth of food supply for the fallacious arithmetical ratio.' In recent years the principle of diminishing returns has been explained thus (p.31 Carr-Saunders): 'In any country under any given conditions there is an optimum density of population which, if attained and not exceeded, will obtain the largest income per head that is within reach.' This is in accord with the general principle of diminishing returns (Benham *Economics* p.128 3rd edn): 'As the proportion of one factor in a combination of factors is increased, after a point, the marginal and average product of that factor will diminish.'

The conceptual categories that were implicit in Malthus's theory are stated explicitly in this analysis. Firstly, that man is an animal and this animal nature remains as an ever present constraint upon each individual despite all rational attempts to control and refine it. On the other hand, man's rational faculties are undoubtedly of an entirely higher order than those of any other animal.

At this point it is useful to add some opinions from the original sources of psychology; the general theme being the refinement of the biological urges or instincts to serve man's more rational abilities:

Freud argued that the development of man's civilisation has been due to the steady sublimation of the sex instinct. Reuben Osborn in *Marxism and Psychology* (pp.63-66) observes: 'He (Freud) said (quoting a philologist who arrived independently of psychoanalysis at these conclusions) that the first sounds uttered were a means of communication, and of summoning the sexual partner, and that in later development the elements of speech were used as an accompaniment to the different kinds of work carried on by primitive man. This work was performed by associated efforts, to the sound of rhythmically repeated utterances, the effect of which was to transfer a sexual interest to the work . . . . We believe, said Freud, that civilisation has been built up, under the pressure of the struggle for existence, by sacrifices in gratification of the primitive impulses . . . . At bottom society's motive (for restraining the instinctive life) is economic; since it has not means enough to support life for its members without work on their part, it must see to it that the number of these members is restricted and their energies directed away from sexual activities on to their work — the eternal primordial struggle for existence, therefore persisting to the present day . . . . We can only guess as to the process by which man built up an effective social organisation with which to limit and redirect his instinctive life. We can be sure that this was a necessary stage in the transition from animal to human life. The theories of Freud and Marx concerning this transition carry conviction, I think, because they present, on a large historical scale, what we can observe going on in the child growing into a social being. He has to learn to restrain his impulses, to limit his demands to fit in with the pattern of social life, to forego many personal ends that threaten social goals. This is the great merit of the Freudo-Marxist picture. It gives an account of the trials and difficulties of early childhood, both of the human individual and of the human race.'

It appears that there is no final victory in the struggle for man's civilisation; it is renewed and repeated from generation to generation; each new birth brings an infant requiring to be socialised from a primitive state into the ways of the current civilisation.

This process of redirecting instinctive impulses is, in my view, most logically regarded as a redirection of primitive biological energies. It is significant that although many different instincts have been, in the past postulated, nevertheless, Freud only recognised two, namely the life instinct and the death instinct. In my view, only the former of these two is a positive and creative refinement of energy, since the latter is entirely destructive. The life instinct is the basis of man's rational self and also of his social and co-operative activities. The death instinct is the source of conflict and derives from the struggle for survival which is the lower animal mode of evolution; this mode of conflicting human activity can also pollute man's rational behaviour to the point where it becomes diabolical animal cunning in the service of war and destruction. It will be demonstrated, at some length, in the later chapter on rationality, that man's higher rational mode of activity must aim at conservation of resources and must avoid conflict and its consequent wastage of resources.

Reuben Osborn observes (p.26):

'Man's rational self, the product of interaction between the strivings of the id for unconditional gratification and the exigencies of the outer world, holds the hope that he will be able to master those inner destructive forces and turn them to socially valuable ends. Freud himself had to rebuke those psychoanalysts who took too despairing a view of the possibilities of rational control and pointed out that, however weak the ego was in relation to the "daemonic forces within us", the growth of knowledge and understanding about human psychology provided the best means of freeing the ego from its enslavement to the purposes of the id. "Where id was, there shall ego be", he wrote. We may therefore look upon Freud's dark picture of the forces of destruction within us urging the world to mutual destruction as a warning of what could be if our rational selves do not take command.'

Of course, the precise manner in which the human brain functions is not known for certain and much research will no doubt be done in this vital area of knowledge; however, it appears reasonable to postulate that the two different modes of human behaviour are originated in two of the different main divisions of the brain; that is to say, the higher rational mode is controlled from the cerebral cortex and the lower animal or instinctive is motivated from the limbic system. The former controls man's most distinctively rational behaviour, whereas the latter is concerned with sequential activities such as feeding, attacking, fleeing from danger and mating — kinds of activities that have often been called 'instinctive'.

The construction of the human brain will be considered in the next section of this book.

It is very significant that much original work in psychology centres upon the concepts of fundamental biological energy and its transference to useful work. Thouless in his book *General and Social Psychology* reserved the term 'instinct' for the energy behind the action; thus he defined instinct as the energy leading to a purposive course of activities common to all members of a given species organism, and having its origin in an innate disposition in the organism.

Jung gives a possible explanation of the way in which fundamental biological energy is converted to useful work. (pp.19-21 *An Introduction to Jung's Psychology* by Frieda Fordham) 'Libido is natural energy, and first and foremost serves the purposes of life, but a certain amount in excess of what is needed for instinctive ends can be converted into productive work and used for cultural purposes. This direction of energy becomes initially possible by transferring it to something similar in nature to the object of instinctive nature. The transfer cannot, however, be made by a simple act of will, but is achieved in a roundabout way. After a period of gestation in the unconscious a symbol is produced which can attract the libido, and also serve as a channel diverting its natural flow. The symbol is never thought out but comes usually as a revelation or intuition, often appearing in a dream.

'As an example of this transfer of energy from an instinctive to a cultural purpose, Jung cites the spring ceremonial of the primitive Watschandis, who dig a hole in the earth, surround it with bushes in imitation of the female genitals, and dance round it holding their spears in front to simulate an erect penis . . . . During the ceremony none of the participants is allowed to look at a woman . . . . they are sharing in a magical act, the fertilisation of the Earth woman, and other women are kept out of the way so that the libido shall not flow into ordinary sexuality. The hole in the earth is not just a substitute for female genitals, but a symbol representing the idea of the Earth woman who is to be fertilised, and is the symbol which transmutes the libido.

'We should note here that throughout his work Jung uses the word "symbol" in a definite way, making a distinction between "symbol" and "sign": a sign is a substitute for, or representation of the real thing, while a symbol carries a wider meaning and expresses a psychic fact which cannot be formulated more exactly. The Watschandis' hole in the earth can be looked on as a representation of a woman's genitals, but it also carries a deeper meaning; it is more than a sign, it is also a symbol.

'There is a very close association between sexuality and the tilling of the earth among primitive people, while many other great undertakings, such as hunting, fishing, making war, etc., are prepared for with dances and magical ceremonies which clearly have the aim of leading the libido over into the necessary activity. The detail with which such ceremonies are carried out shows how much is needed to divert the natural energy from its course. This transmutation of libido through symbols says Jung, has been going on since the dawn of civilisation, and is due to some thing very deeply rooted in human nature. In the course of time we have succeeded in detaching a certain proportion of energy from instinct and have also developed the will, but it is less powerful than we like to believe, and we still have the need of the transmuting power of the symbol. Jung sometimes calls this the "transcendant function".

'Jung's view of the unconscious is more positive than that which merely sees it as the repository of everything objectionable, everything infantile — even animal — in ourselves, all that we want to forget. These things, it is true, have become unconscious, and much that emerges into consciousness is chaotic and unformed, but the unconscious is the matrix of consciousness, and in it are to be found the germs of new possibilities of life. The conscious aspect of the psyche might be compared to an island rising from the sea — we only see the part above the water, but a much vaster unknown realm spreads below, and this could be likened to the unconscious.'

These views of Jung bear out that man's animal nature and his higher reasoning faculties are very intimately interrelated. Nevertheless, that is not a good reason for failing to recognise the significance, for analytical purposes, of the fact that man has evolved from and still incorporates an animal body and brain; and that the higher rational faculties are entirely controlled by the cerebral cortex, a complex additional controlling brain which has evolved around the older animal brain. Analysis requires that the effects of these two brains, the older and the newer in order of evolution, shall be considered; at present, the interconnection between these two brains, or parts of the human brain, and the method of functioning may be imperfectly known; however, the existence of two parts of the brain points the problem of social science: which will control, the lower animal or the higher rational? We know which should control but in practice the ideal does not always happen. Often it may be that the lower animal, instinctive part of the brain is dictating to the higher rational part and indeed polluting it to the level of the lower animal evolution and conflict for mere survival.

Jung sees these interrelationships as associated with a collective consciousness; Fordham p.27 states: 'The unconscious therefore, in Jung's view, is not merely a cellar where man dumps his rubbish, but the source of consciousness and of the creative and destructive spirit of mankind.' And also: (p.24) 'We may hazard a guess that the primordial images, or archetypes, formed themselves during the thousands of years when the human brain and human consciousness were emerging from an animal state.

'Jung does not mean to imply by this that experience as such is inherited, but rather that the brain itself has been shaped and influenced by the remote experiences of mankind. But although our inheritance consists in physiological paths, it was nevertheless mental processes in our ancestors that traced these paths. If they came to consciousness again in the individual, they can do so only in the form of other mental processes: and although these processes can become conscious only through individual experience and consequently appear as individual acquisitions, they are nevertheless pre-existent traces which are merely "filled out" by the individual experience. Probably every impressive experience is just such a break-through into an old previously unconscious river-bed.'

It is not the intention in this volume to attempt any full appraisal of the psychological literature which bears upon the relationship between man's animal being and his more rational faculties. However, it is important to insist that the study of such relationships is a very valid and vital aspect of social science.

10

EVIDENCE FOR THE TWO MODES OF EVOLUTION FROM THE CONSTRUCTION OF MAN'S BRAIN

The account given in *Introduction to Psychology* by Hilgard and Atkinson, 4th edn is a useful summary of the hierarchical organisation of the human brain: (p.40) 'The human brain consists essentially of the structures of the primitive vertebrate brain preserved in similar anatomical relations but of course modified in detail, plus the enormously developed cerebral cortex built upon the older brain. It helps to think of the human brain as composed of three concentric layers: a primitive core, an older brain evolved upon this core, and an outer layer of new brain evolved in turn upon the second layer. All three layers are, of course, interconnected in a complex fashion, and a new layer cannot evolve without changing the conduction of impulses to and from the earlier and more central layers of the brain.' (Pribam, 1960)

(p.41) 'The central core . . . . functions in such life-maintaining processes as respiration and metabolism, in the regulation of the endocrine gland activity, and in maintaining homeostasis . . . . We thus see how this central core of the brain provides for the very important life-maintaining functions.' . . . . 'The Old Brain: Around this central core are the older parts of the brain that serve somewhat more complex functions . . . . These structures are commonly known as the limbic system . . . . Pribam (1958) suggests that if all the data are combined, we would find the limbic system is concerned with sequential activities, that is with activities that proceed for some time and involve a number of movements before they are completed. These include the activities of feeding, attacking, fleeing from danger, mating; kinds of activities that have often been called "instinctive" . . . . The evidence from both lower animals and human studies suggests that the limbic system builds upon the homeostatic mechanisms, regulating the dispositions of the organism to engage in sequences of activities related to the basic adaptive functions mentioned above.'

(p.44) 'The two large hemispheres at the top of the human brain represent man's "new brain". The convoluted layer of gray matter — the cerebral cortex — that covers them controls man's most distinctively human behavior.' This includes the motor area, the body sense area, the speech area, the association area that must serve to bring together phenomena involving more than one sense and must be involved in learning, memory and thinking. Other parts, such as the forebrain, may serve kinds of memory that the limbic system does not serve. Still other parts serve for learning and problem-solving and, in man for language. (The foregoing passage is a summary of pp.44-48)

(p.42) 'By comparative study we find that parts of man's brain are very similar to the brains of lower animals. The parallels found in the structure between man's brain and simpler brains lead us to infer similar functions, which need to be verified by further investigation. Another way to study the evolution of the brain is through observation of its embryological development, for in its early stages of development the brain of the human embryo reveals its relation to the brains of organisms lower in the evolutionary scale.'

However, (p.42) 'treating the brain in this way, as three concentric structures — a central core, an old brain, and an outer core — must not lead us to think of these interrelated structures as being independent. We might use the analogy of a bank of interrelated computers. Each has specialised functions, but they still work together to produce the most effective results. Similarly the analysis of information coming from the senses requires one kind of computation and decision process, for which the cortex is well adapted, differing from that which maintains a sequence of activities (limbic system). . . . All these activities are ordered into complex subordinate and superordinate systems which maintain the integrity of the organism.'

Therefore, to summarise briefly, it is vitally important in social studies to set up a theory or model incorporating the two different but interconnected modes of behaviour or evolution. Not only does the conflict and co-operation in human affairs indicate this, but also the very construction of man's brain is consistent with man having a higher brain superposed upon a lower one; the former being the source of distinctively human activity and the latter being more akin to the rest of the animal kingdom.

The work of Malthus included a concept of man's population growth which can be interpreted as an 'ideal type' of animal motivation or evolution. He also suggested that certain rational controls of population should be made, as for example by later marriage and abstaining from sexual intercourse; however, he did not anticipate the growth of artificial birth control to the extent which has subsequently taken place.

It is not a particularly telling criticism to point out that the two modes of activity are not separate but interrelated. It is important to recognise and define them in order to investigate their relationships.

Clearly a purely animal mode of human behaviour did not exist in Malthus's time since man was already established as a uniquely intelligent species with many great achievements and inventions to his credit. However, the progress made in rational innovation has been even more remarkable and dramatic in the subsequent years; and this rationality has been the major characteristic which distinguishes man from the rest of the animal kingdom.

By a process of logic and insight, Malthus appears to have set up as a datum an animal mode of population growth which conceptually might have existed in man's immediately prehominoid state at the outset of his mutation into the new species, man. The next chapter will give a very brief résumé of the ways in which man has developed since Malthus's time with tremendously accelerated progress along the path of increasing rationality; and in ability to organise and plan his affairs by understanding and controlling his environment.

A word of caution is, however necessary here, to the effect that despite his great achievements, the veneer of his civilisation still remains very thin. This is so because he still retains the same biological animal constitution as he had many thousands of years ago. Therefore, the whole edifice of civilisation could easily crumble into conflict and warfare which is rendered more deadly by man's potential for diabolical cunning.

Carr-Saunders points out that from early times man maintained small families by means of abortion, infanticide, or sexual abstention. This is very different from the situation of other animals (and presumably also of man's prehominoid ancestors) which have none of these checks on population. Such population checks, even though they may be only customary, nevertheless entail some thoughtful, human, content. It has been, of course, only during the most recent advances in science and education that man has begun to develop his fuller rational faculties; however, the distinctive ability of man's thinking has been, for thousands of years, clearly established as of a higher order than those of other species. Thus he managed to maintain a difference between his fecundity and his fertility such that for long periods of man's history his population has been fairly stable.

It is quite fair to point out that the checks used had some rational content even though they were reinforced by customs, emotions, and even magical rituals. As Carr-Saunders suggests (p.35 *Population*):

'If these customs are beneficial, those who practise them will be selected. And surely these customs are beneficial if they enable those who practise them to maintain numbers at, or close to, the desirable density.'

11

A BRIEF RÉSUMÉ OF MAN'S RATIONAL PROGRESS SINCE MALTHUS

Carr-Saunders observes (p.42 *Population*) 'While it is impossible to estimate the prevalence of contraceptive practice and of abstention from intercourse, it is probable that they account for the whole of the decline (in birth-rate) which the figures show.'

Therefore the very substantial decline in birth-rate that occurred after 1875 was probably due to some degree of thoughtful interference with natural trends in birth-rate: moreover, the even greater fall in the death-rate over the nineteenth century was undoubtedly due to the rational advances in sanitation and medicine.

There have been many inventions and innovations in both agriculture and industry bearing witness to increasing rationality in man's activities. Agricultural innovations included the introduction of new crops and rotations of crops, manures, fertilisers, reapers, harvesters, the drill, seeder and a great variety of mechanical devices; these have yielded a considerable growth in knowledge, techniques and productivity in food production since Malthus's time.

Scientific advances in steam, gas and petrol engines enabled food to be conveyed more cheaply and thereby removed the necessity for local self-sufficiency. Science in agriculture and transport by steamship and rail helped to open up the prairies and South American wheat lands for large-scale food production: while rational scientific and industrial progress enabled Britain to increase rapidly her exports of manufactures so as to pay for the increasing food imports. This enabled the feeding of populations that increased despite the fall in birth-rate and because of the rapid decline in death-rate.

Therefore, from 1800 to 1875 was, in Britain, according to Carr-Saunders, (p.41 *Population*) 'in general a period of almost unrestricted multiplication, and as such an almost, if not quite, unique epoch in the history of the human race.' Thus there was great migration from the country estates which could not produce enough to support an increase in which families of ten or more were quite usual; meanwhile, the new manufacturing industries grew rapidly in the towns to absorb the displaced labourers; and thus enabled Britain to exchange manufactures for food in world markets.

In the rapidly expanding towns more rational planning in power industries, sanitation and civic services grew: education was improved to meet the growing demand for skilled artisans clerks and other personnel in the developing industrial concerns and to raise the effective use of intelligence by the whole population during a period when the franchise and popular democracy were also being extended greatly. Over the whole period a very large improvement in general living standards also took place. There were changes in the methods of dealing with problems of poverty during the period since 1800.

The Speenhamland system of outdoor relief had kept many people out of the workhouse; but this, it was claimed, had led to indiscriminate growth of families and lack of responsibility on the part of many labourers as well as employers. The Poor Law Amendment Act of 1834 was regarded by many as a punitive measure and deterrent to irresponsibility; the workhouse test was reimposed thus causing the break-up of many families through ill-health, incapacity or misfortune.

This system continued, with some modification until 1909 when a Royal Commission decided that it was not an effective deterrent to improvidence; and it was discontinued during a period of social reform which led eventually to the welfare state as we know it today. That is to say that after 1834 the poor were treated rather like irresponsible animals but, by a process of development, today poverty has come to be viewed in a much more humane and rational context.

The whole picture is one of increasing use of science, education, rationality; and the growing replacement of muscle-power by exploitation of coal, oil, gas and other sources of natural energy.

However, although we can make general claims to increasing rationality, nevertheless, it is also necessary to define much more precisely what is meant by rationality; this is essential for the logical development of social analysis. Unfortunately it has been quite usual for social theorists to use terms like rationality and efficiency without making any serious attempt to define these terms. This has been the source of much confusion.

On the other hand it is equally confusing that social theorists fail to state explicitly the facts concerning man's two interwoven modes of behaviour. Such a model needs to be clearly described using precise definition of rationality; since this is the outstanding factor differentiating between the two modes of human activity. Therefore the next chapter will outline a detailed definition of rationality.

12

RATIONALITY

It has been observed in an earlier chapter that over-simplification of the term rationality can cause much confusion and misunderstanding. However some attempts have been made to compensate for an unduly simplified concept of rationality by postulating different kinds of rationality. Max Weber distinguished between formal and substantive rationality. He defined formal rationality as the translation of all situations and decisions into numerically calculable terms or their subsumption under technical rules: while substantive rationality, he defined as, weighing the goals and means of action against various value scales such as justice, power, taste, etc.

My own view is quite contrary to such conceptual divisions in rationality; if something is good, by any valid standard, then it is also rational: it is quite possible for people to be unaware of some essential ingredients which would establish that rationality, but that does not justify the postulation of numerous different standards of rationality, in the way that Weber suggests.

Weber's concept of substantive rationality can encompass an infinite variety of different goals and value scales, as for example, power to wage war; this does not, in my view, represent a distinct form of rationality, but rather it is a misdirection or corruption of man's rationality to serve a useless and wasteful aggressive instinct. There cannot be a war in which both sides are fighting a 'just war': rationality of objective can only exist in those who strive to preserve justice. Those who wage war for purposes of domination and power, may be rational in their methods but in their objectives, they are corrupt barbaric and irrational. Of course, it must be conceded that in many wars the motives of both sides may be, to some extent corrupt; nevertheless, it is important to emphasise that there must be a common basis for both the processes and the objectives, if they are both to be accepted as rational in any enterprise.

Since there seems no point in accepting a variety of different criteria or definitions for rationality, it is therefore necessary to consider a single, more detailed definition. In this context it is necessary to discuss the whole process whereby rational human life has evolved from inanimate matter. In fact it is more instructive, initially, to consider constructing a model of the human rational process in an evolutionary context.

At the lowest level of inert matter the functioning of the laws of C.O.E. are apparently quite automatic. Living organisms are constrained to take up and use this inert matter. In general, apart from man, these living organisms do not have rational scientific understanding of natural laws of matter in any way similar to that devised by man; however, they do possess certain instinctive, reflexive or tropistic reactions to stimuli which enable them to survive: that is to say, they have dynamic equilibrium or homeostasis. This depends, essentially upon maintaining a suitable balance of internal energy and resources; and also upon maintaining an adequate communication and contact with their environment for survival. This is not generally recognisable as rational intellect as we know it in man, but it often suffices for survival. However, some organisms and species do pass out of existence because they cannot maintain this dynamic equilibrium.

By contrast man is not equipped biologically with weapons like claws and tusks for attack or thick shell for defence or great speed for flight. Nevertheless, he is very adaptable. Man is an animal with animal instincts and reflexes; his brain has a part which is quite similar to a corresponding part of the brain in other animals. Nevertheless, the cerebral cortex in man is much more highly developed; thus it is able to control or sublimate more of the cruder animal impulses or energy into more carefully thought out channels.

Man is able to study, scientifically, and understand how to control the inert materials and resources, how these take part in the systems of living things and in his own body. Of course, men do not achieve such understanding and control to perfection; but it is the ability of man to set up such ideals that elevates him above the other animals; and this is a most vitally important part of any realistic model or definition of human rationality, that he can strive towards perfection in understanding and control of himself and his environment.

It is also true that men are not all of equal intellect and ability; thus in human society there is usually some organisation and authority whereby it is attempted to arrange that those of higher rational attainment superpose their expertise upon the efforts of the less able.

Therefore, it appears that social groupings of men in co-operation are of potentially much higher rational attainment in activity than are individuals; since they can utilise the highest intellects in conjunction with the economies of division of labour on large scale. This is again, an ideal from which man in society falls far short, in so far that the vitally important requirements of mutual regard and co-operation are often lacking; whereas the urge to dominate and exploit are greatly in evidence. Thus the higher rational mode of rational co-operation is often submerged beneath the lower animal urges towards aggression, conflict and domination.

In the foregoing remarks there have been identified, four kinds of system, built up in hierarchy one upon the other, ascending in order of complexity and 'rationality' of their controlling networks; each higher system builds itself by using the lower systems as its bricks or raw materials.

The lowest is the inert which is purely mechanistic and apparently lacking in rationality; the next higher kind of system is the living organism; some of these possess a sort of instinctive rationality of a low order. The next higher system in the hierarchy is the individual man with a generally much higher rational potential; and the highest possible potential in rationality can only be envisaged in the system at the very pinnacle of the evolutionary hierarchy, namely 'man in society'.

When we view this evolutionary process and attempt to select from it that which is rational in model or definition form, then we produce, not a whole science of man, but simply one abstracted concept. It is a concept representing some thing that never exists in isolated or pure form; it is in a word, an 'ideal'. Nevertheless, this ideal is a fact of human thought, and, as a carefully defined concept, it is also a fact of social science.

At this point, it is appropriate to emphasise and remind ourselves of a vitally significant attribute of man: the one characteristic that raises him above the other animals:

Man is capable of envisaging ideals and is also able to strive towards them with 'heart and mind' that is; the lower animal part of his brain can become attuned to and controlled by the higher rational cerebral cortex, so that the biological energies are refined and sublimated into higher channels. This involves a harmony of mental orientation and a feeling of desire for mutual co-operation and regard which is the basis of creative social co-operation.

This view of ideals is not incompatible with Freud's opinion that civilisation is due to the steady sublimation of the sex instinct; however, it does suggest something rather more than just that. On the other hand, my views, as expressed here could be fairly criticised as a scientific parody of a subject that needs to be very much more profoundly studied.

Bearing in mind, therefore that rationality, in any pure form, is an ideal, and assuming that any variability of activity due to a controlling nervous system is to some degree rational; the systems model for rationality can be summarised as follows:

There are four easily identified levels of evolution, the inert mechanistic, the animal biological level, the level of individual man, and man in society. These four levels rise in order of increasing rationality, decreasing determinism, and increasing variability.

In order to understand and control his environment man must apply rationality to understanding the systems at each of these levels. In every one of these systems, the balance of energy and resources is of vital importance: and one must have in mind a model of systems with energy balances built up using systems of lower order of rationality as their raw material.

This model of rationality sees man as controlling his environment through his understanding of natural laws, and particularly by understanding of the laws of C.O.E. and the concept of the energy balance in relation to dynamic self-maintenance. This requires further discussion; but before proceeding in that direction, it is worth noting that the model which has been proposed for rationality effectively disposes of some rather troublesome objections that are often made against concepts of rationality: firstly it is objected that men are rarely if ever rational; secondly it argued that rationality is an unattainable ideal; thirdly it is pointed out that some men are much less rational than others; fourthly it is often erroneously suggested that conflict alone is the motivating influence in evolution, (thereby neglecting rational co-operation); and fifthly, it is often pointed out that other animals are also intelligent. A careful consideration of the proposed model of rationality will put all of these objections into a proper perspective.

In answer to these objections it can be emphasised that rationality is all about setting up plans, goals, or ideals and striving to attain them, however imperfectly; that in this process it helps if the more intelligent guide those with lesser capacity; that when evolution rises above the mechanistic level, and enters the rational sphere, its progress in rationality can be judged by the increase in co-operation and the resultant reduction in conflict. And that finally the model does admit of intelligence in animals other than man; but it is quite certainly of a lesser order.

At this point it is now convenient to return to the question of substantive rationality. My contention is simply that man's goals, objectives and ideals should be guided by the same basic rational processes as those that guide the detailed pursuit of those objectives; it is the same set of natural laws that need to be considered and satisfied; the laws of C.O.E. still apply and an energy balance still needs to be maintained. Any objective which causes the deterioration and undermining of the social system without providing for compensatory change and modification to the system is irrational; that is, its objectives should be consistent with dynamic self-maintenance of the social system (homeostasis); or alternatively changes in form and structure of the social system must enable a new dynamic equilibrium to be established (defined by Walter Buckley as 'morphogenesis').

The kind of activities that are clearly irrational under the concept of rationality outlined above are: profligate waste of resources, pollution of land, water and air, and destruction by warfare.

Of course it makes some selfish sort of sense to grab all the available resources thus depriving other people, to pollute the water leaving it impure for those downstream and to dispossess neighbouring peoples by defeating them in war or by economic manipulations. However from the viewpoint of mankind as a whole, these achievements are extremely wasteful of human resources, as well as being both unjust and irrational. Such wastes arise from the inability of men to find grounds of co-operation and coexistence in agreed rational ways: thus, vital issues which require co-operation are put to the wasteful test of bloody conflict: and man's crowning rational attribute is perverted to serve antagonism and destruction. This serves to confirm the still widespread persistence of animal modes of conflict and evolution in the affairs of men: moreover it is beyond the wit of one man or one nation, however virtuous, to remedy these defects. This conflict is not rationality; but it is the environment into which our model of rationality must fit.

Such considerations lead us on to discussion of the central dilemma of human rationality: which is, briefly, whether if man were to attain perfect rationality, his society would function as well-oiled clockwork performs its predetermined paths and revolutions; and would the men act as identical cogs in the gigantic wheels and mechanisms of social bureaucracy? If that were so, of what value would all his striving for rationality be?

If the model described in this chapter is realistic, then such a deterministic nightmare need not concern us; since the whole basis of this model is of evolution building up from the invariable, mechanistic, inert matter to the variable, rational, human living social system. Rationality clearly increases with variability of activity.

Much of the difficulty in the concept of human rationality arises because of this change from deterministic to rational mode of evolution; and the problem arises because the change has to be attempted in the form of an abrupt step in an upward direction which man must either make or accept the consequences of failing to make it. If man fails to make the desirable advances in rational social co-operation, then he will continue to use his superior mental attributes for the purposes of conflict.

In fact it must be admitted, at this moment, that despite some impressive appearances of rational social co-operation, nevertheless, there is also equally impressive evidence of failure in that direction. This failure is due to deep rooted, inbuilt habits or instincts of aggression and conflict; these arise from the situation whereby evolution has proceeded from the activities of particles and very limited systems which are only capable of very limited reactions to their environment.

As evolution has proceeded the units have gown more complex and their variability of reaction to their environment has also increased until, in man, it has become very great in its potential. It is this great potential for variability and rationality that precipitates the crisis and the need for a new wider outlook.

In this wider viewpoint both the question of survival and also the question of individual satisfactions are bound up with the necessity for a wider view, a viewpoint in which the interests of mankind as a corporate whole must be considered above limited sectional interests. This has now become essential because the power of sectional interests for conflict has now become so great that they are now capable of destroying utterly the whole fabric of human civilisation and, indeed life itself. The original animal methods of evolution, proliferation, conflict, and struggle for survival, are no longer appropriate these days, when conflict may leave nothing! It is in many respects unfortunate that man retains, in his brain structure, the instincts and mental dispositions of the animals from which he has evolved. His only hope lies in his higher brain faculties and the wider view of humanity as a whole.

On the other hand there is much more satisfaction to be

gained from this wider view than that of mere survival. Indeed, the wider view of humanity needs to be much more firmly based than in the miserable fear of destruction; and there are, certainly some very positive higher motives for this wider view. In particular, there are the motives of mutual regard, caring, belonging and the search for truth for its own sake; this list is not exhaustive, but it will serve.

However, if we could only have the necessary enlightened spirit, there would also be the benefits of great inventions without the wastage and travail of warfare. If we could obtain full co-operation between men the material benefits in goods and services would proliferate enormously; thus solving the problems of scarcity of resources by the economies of division of labour.

Alas, these are ideals; and we are very far from achieving them at this moment, nevertheless, they are the ideals of rationality; and if they were achievable, there could be a great variety of different cultures all fitting together happily in appropriate niches without any need for standardisation either of individual men or of communities. It would be meaningless to calculate or speculate as to which type of society or culture might be the most efficient since they would be complementary.

Therefore, it is important for men to examine their motives. As animals, men have grown accustomed to striving for material gain for the sake of survival and such self-centred behaviour has become habitual although the necessity for it has gone for many men. For many, the aim of life is no longer mere survival. Therefore, it is now essential that man should modify this deep rooted obsession with individual material gain and struggle for survival or domination which derives from his origins in animal evolution.

The great challenge today lies in the wider concept of man in society; in this man must strive for harmonious relationships on a world scale. He must strive to eliminate the enormous wastes of the arms race, the bitterness and conflict due to race hatred and national antagonisms, the poverty of the underdeveloped nations, and the greed, fear and aggression which motivate so much of man's activity.

It has been pointed out in numerous United Nations reports on world social problems, that their solution depends primarily upon the development of education to combat poverty; that is to say education, initially at the most rudimentary level. However, we must not be deluded into the belief that educating the underdeveloped nations to the standards of western capitalism, communism or any of their intermediate brands of political system, will solve the problems of mankind. Are not these very systems the basis of world conflict? And are they not poised day and night ceaselessly ready to deliver destruction upon each other? And at what cost in outpourings of man's misdirected intellect and resources is this fragile balance of fear maintained? The case for conservation of man's energies and resources does not need to be argued, for it is self-evident.

What is now required, most urgently, is a scientific social theory upon which all can agree and rely, a theory that can be the basis for an ideal of co-operation on a world scale and upon conservation of man's energies.

It is the obsession with individual gain that lies at the root of all conflict; therefore it must be abandoned and men must seek their satisfactions in the intrinsic qualities and patterns of social relationships and culture rather than in the accumulation of possessions and power.

To summarise briefly: Given co-operation and mutual regard between men, and a search for the truth, for its own sake; then the extension of his control of his environment and the proliferation of his material goods and resources would follow automatically. Therefore, there would be no need to strive for some pernicious efficiency in uniformity or optimum design of culture; because the material resources derived by co-operation would be immeasurably greater than the parsimonious cheese-parings of such misdirected optimisations. The resources required for variability of culture would arise as by-products from co-operation and mutual regard.

13

DEVELOPMENT AND DISCUSSION OF THE HYPOTHESIS

The principles of C.O.E. and uncertainty are interwoven into man's activity. Once resources have been committed and used in one chosen way, they cannot be used again in some other way, unless a recycling process is employed. This is so because energy cannot be created or destroyed: and matter is, by Einstein's theory, simply one particular form of energy. It is the great generality of the phenomenon of energy that makes it of such importance to human problems.

The bodies of men and their whole environment are pervaded by this entity that is scientifically called energy. Without in any way implying that energy is an ultimate 'reality', it is, nevertheless, quite justifiable to use it as a basis for scientific analysis. It is worth noting, in this context, that this scarcity of resources together with an assumed insatiability of human wants, also constitutes the basis of economic analysis, one of the most developed of the social sciences.

A man may choose how he will dispose his resources; but they are limited; he cannot run a two-minute mile; he might only manage a four-minute mile if he denies himself many things and conserves his energies carefully to the purpose. He may choose to leap from a high building, but will probably die as a consequence. He may choose not to work, and though he may not starve, he will not be likely to live in the greatest of comfort. He may prefer to use out-of-date techniques of industry, but he may well find that he earns less than those who take advantage of newer methods. In brief, he may choose how he allocates his resources; but he must face the consequences, because those resources are limited; and once committed cannot be used again without recycling.

In this way the principles of C.O.E. and of uncertainty are interwoven in the social sciences to produce the further principle of man being constrained to accept the consequences of his choice of actions. Thus in modern social geography the cruder forms of physical determinism are no longer widely accepted: but there has developed the concept of possibilism. This implies that in many areas man is presented by nature with a range of possibilities within which he may choose how he fashions his social economy and culture. This is regarded by some as a qualified sort of determinism which, in my view, consists of being constrained to accept the consequences of one's choices of action.

Therefore, men are able, to some extent, to assess the way the world about them functions, though not always quite accurately; and how their bodies fit into that environment. The general laws that man has, so far, discovered are principally those of energy; the more fully he understands and discovers such laws, the better can he turn such knowledge to the purpose of satisfying his desires. If he fails to understand these laws or chooses to ignore them, he does not thereby suspend their operation; they are the medium within which his physical activities must be carried out. He cannot escape the consequences of the choices of action that he makes.

The change from the domestic system to factory production in the late eighteenth and early nineteenth centuries well illustrates the operation of the laws of energy; and the way in which men take the consequences of their choices of action.

In the second half of the eighteenth century it became clear that by developing machinery for spinning and weaving by means of power-driven devices, greatly increased production could be obtained. Some inventors may have worked for the pleasure of creating something new; or they may have been interested in being able to produce more cloth on their own small scale, having little regard for the possible effects of their inventions upon the industry as a whole. Samuel Crompton who invented a spinning mule, appeared to be of this temperament.

However, Richard Arkwright saw the immense possibilities in increased production, wealth and influence to be achieved by harnessing these inventions under one roof in a factory using water power. Some machines he invented, some he improved and others he cribbed; thus he fathered the new factory industry based upon water power with great success and increase to his power and wealth.

Meanwhile, the domestic spinners and weavers began to find their earnings drastically reduced by competition from the increasing quantities of factory produced cloth. They did not very willingly accept the factory system, in fact there were outbreaks of machine breaking. However, many domestic workers soon found themselves reduced to poverty through persisting with the less efficient domestic methods. Large numbers chose to work as wage-earners in the factories; and eventually the old techniques failed to produce either the quantity or the quality to yield even a meagre living.

No doubt the above account is very over-simplified; but it does indicate that there was nothing automatic or deterministic in the process of changing to factory production. The people involved were able to choose and were constrained to accept the consequences of their decisions; some went to the factories; some became very poor; others starved and Arkwright grew rich and powerful. All these things happened because of innovations which harnessed water and later steam power for spinning and weaving. Of course, it can be argued that there was economic pressure or compulsion. Nevertheless the inventions and their application to factory production required rational thought and decisions as also did the acceptance of factory authority and methods; they were by no means automatic or predetermined. It is the whole purpose of this illustration to show that such economic pressures were the result of man, by rational thought harnessing sources of natural energy for the purpose of producing things that he required.

At this point it is convenient to attempt a further detailed discussion of the general hypothesis: If we pursue the theme of the foregoing rather crude historical analysis, it would appear that man has the potential to increase his productive capacity by acquiring control over increasing sources of energy or by using the energy which he now controls more efficiently.

However, this would place too great an emphasis upon the purely material aspects of man's potential. Nevertheless, it is probably not too inaccurate a summary of what men are striving for most of the time. The motivation as thus described is material gain; and as such it partakes of superior animal cunning rather than of higher rational activity.

In fact the above statements assume the objectives to be agreed; productive capacity is referred to as through the actual commodities and satisfactions obtained, were never in doubt. Such statements deal with what Weber called 'formal rationality'; and they neglect what he termed 'substantive rationality'. Such an approach is far too tolerant of divisive sectional interests, their selfish objectives and the pursuit of individual material gain for its own sake; while such a viewpoint neglects, catastrophically, the wider viewpoint of man in society; and the urgent need, on all sides, for the elimination of the wastage due to conflict.

A much more careful and fundamental appraisal is required, if we are to appreciate man's potential for higher rational ideals. This has been discussed, at length, in the previous chapter on rationality; and it was concluded that the higher rational mode of human action is bound up with co-operation and a spirit of mutual regard; whereas, the lower animal mode is characterised by conflict and an attitude of antagonism.

Considering, therefore, man's higher rational potential; he has the capacity, unique among animals, increasingly to understand and control his environment. Now this implies each of the four levels which were defined in the previous chapter on rationality. That is, he must seek understanding and control at the levels of, firstly, inert matter, secondly, living organisms, thirdly, individual man and fourthly at the all-important level of man in society.

It must, at this point, be emphasised, that for higher rational action men must make a very genuine effort to understand that major part of his environment which consists of his fellow human beings, that is 'man in society'. In this context, the spirit of mutual regard and co-operation must be made the basis for a creative unity of purpose in man's social activity. In particular, the objectives set and manner of satisfying wants must be very carefully considered in order to ensure that one person's or group's satisfaction is not obtained at the cost of undue hardship or injustice to other people.

Having examined the hypothesis carefully, it is now logical to consider the various innovations, inventions or devices which men have conceived and made in order to harness natural sources of energy for his own purposes. Such exploitation of energy is certainly one characteristic of man which distinguishes him from other animals; nevertheless, this is only a distinction of degree rather than of kind, since it must be acknowledged that these resources may be used for purposes of conflict. Thus man may merely behave as an animal of superior cunning; men may be only fashioning more and more complex weapons with which to eliminate each other in the struggle for animal survival.

Every new invention can be adapted either for peaceful or for warlike purposes; aircraft can carry either passengers or bombs; modern rockets can be used to launch either communications satellites or atomic warheads; computers can be used for innumerable peaceful purposes; but also they may be employed to direct the operations of a third and perhaps final world war: a war that could consummate the ultimate logical conclusion of the lower animal mode of evolution, namely ceaseless conflict leading inexorably to mutual destruction as the intellect and total energy of the process develop and escalate. In the words of Hobbes: *'bellum omnium contra omnes'* and the life of man being 'short, nasty and brutish'.

Therefore, let us proceed to the consideration of these devices for harnessing and conserving energy, in a more thoughtful frame of mind than some theorists have adopted in the past. The devices are, in a sense, themselves, neutral; but the uses to which they are put may be either of co-operation and good or alternatively of conflict and evil. A simple recounting of the history of inventions does no justice to the moral themes, the good and the evil that interweave in human affairs; nevertheless such a history is a relevant and indeed distinctive aspect of man's unique role in evolution; in these inventions and technical achievements he has outdistanced by far the attainments of all the other animals; and has thereby advanced to the point where the ability to discriminate morally is vital to man's very survival.

Having, I hope, quite emphatically, and I hope effectively stressed that the fact of inventing a device for harnessing or conserving energy, does not ensure its being put to good use; it is now time to make a small list of such devices.

14

THE DEVICES AND MEANS FOR HARNESSING AND CONSERVING ENERGY

At this point, only an overall classification is intended and a few examples are given:

Machine Tools and Implements: Saw, Needle, Wheel, Engine.

Means of Communication: Telephone, Radio, Television.

Symbolic Representation: Speech, Writing, Plans, Models, Money.

Methods of Group Organisation: Limited Liability, Authority, Obedience, Specialization.

Intermediate Between Symbolic Representation and Methods of Group Organisation: Rules, Laws, Scientific Methods.

Feelings, Beliefs, Attitudes and Ideals: Mutual Regard, Religion, Scientific View.

Customs and Rituals: Marriage, Parliament, Rituals.

It is not only machines that are necessary for harnessing energy; but also every device that helps to form the social environment in which the machines must operate, is a means whereby energy is harnessed. Moreover, it is worthy of consideration that even the native's spear concentrates his puny bodily energies into a vital spot where it can kill an animal before it kills him; thus it enables him to survive.

The devices for harnessing and controlling energy are further discussed in chapter 19; and their classification is elaborated in the summary p.90.

15

THE HYPOTHESIS DISCUSSED IN RELATION TO THE THEORY OF MARX

There is one aspect in which the two theories are closely similar, namely in the importance attached to technical innovations. In reference to *The Communist Manifesto* of which he and Marx were joint authors Engels stated:

'That the fundamental proposition which forms the nucleus belongs to Marx. That proposition is: that in every historical epoch the prevailing mode of economic production, and exchange and the social organisation necessarily following from it, form the basis upon which is built, and from which alone can be explained the political and intellectual history of that epoch; that consequently the whole history of mankind (since the dissolution of primitive tribal society, holding land in common ownership) has been a history of class struggle, contests between exploiters and exploited, ruling and oppressed classes; that the history of these class struggles form a series of evolution in which, nowadays a stage has been reached where the exploited and oppressed classes (the proletariat) cannot attain its emancipation from the sway of exploiting and ruling class (the bourgeoisie) without, at the same time, and once and for all, emancipating society at large from all exploitation oppression, class distinction and class struggles.' (p.129 *History of Socialism* by Harry W. Laidler)

The above passage contains two ideas that are significant in comparison with the concepts put forward in this book. F. J. Sheed in *Communism and Man* p.21 expresses the first point thus: 'From the beginning of history men have been faced with Nature and their survival has depended on the success with which they could produce from Nature such things as they needed. Production, thus understood, is the primary human activity. History, he says, in so many words, is the history of the economic process. The particular form of production at any given moment — the particular form, that is, in which man bends Nature to supply his needs — determines the whole nature of society.'

This idea is somewhat similar to the concept of man harnessing natural sources of energy by means of inventions and devices. However, one cannot agree that the particular forms of production actually determine the whole nature of society; for in that case, how could it be that Russia and America are the products of very similar technological developments? The key to this difference of viewpoint lies in the precise meaning ascribed to the word 'determine'. Marx used the word determine in the sense of 'sooner or later it is inevitable' rather than as an indication of a precise timetable of events. Nevertheless, there are many people who very strongly doubt that there will be an uprising of the proletariat in America or in Britain which would succeed in establishing Communism.

The view put forward in this book is that the energy harnessing machines, devices and technology are a very powerful constraint, but they are by no means the only influences upon the nature of society. Modern technology and machines can be put to a variety of uses, some good and others quite evil; and the beliefs or ideals of people are often important in deciding which purposes, good or evil, will be served.

In particular, the highly emotional belief in the necessity for class conflict and bloodshed, as expressed by Marx, is an evil influence; it is charged with the animal emotions of hatred and aggression; these he puts forward as the only impulse behind human evolution; and this at a time when the impulse of man's evolution should be turning increasingly from the lower animal mode of conflict and towards the higher rational mode of co-operation and conservation of energy.

Much of Marx's analysis concerning the factual existence of class conflict is perfectly correct, but that is not a good reason for goading men to more conflict when what we clearly need is less! The point is made very emphatically in the present analysis that man has two quite different and co-existent modes of activity and evolution. Marx was clearly very much influenced by the climate of opinion that surrounded the theory of evolution by his contemporary Charles Darwin; this is very briefly summed up as natural selection by struggle for survival. If one applies the same theory to man it is perfectly logical to place great emphasis upon conflict and another great thinker of Marx time, namely Herbert Spencer, was vastly impressed by the significance of Darwin's theory in the field of human activity. The concepts of conflict and competition were also ranked as very important in the theories of the Benthamites or utilitarians who were influential during the life of Marx. I do not suggest that Marx took his ideas directly from these contemporaries; but his emphasis upon conflict was quite in keeping with the current philosophies of his time.

It is fair to say that Marx was certainly an extremist; Spencer, for example, had a great deal to say about co-operation as well as about conflict and competition. However there is an even more important observation to be made at this stage.

The animal mode of evolution as expounded by Darwin has never been satisfactorily applied to man; there were the writers of what Sorokin (*Contemporary Sociological Theories*, ch IV-V) described as the 'bio-organismic school of sociology'; but these did not convincingly show man to be solely influenced by animal evolution, as expounded by Darwin.

It is the particular object of this book to demonstrate that, in addition to the lower animal evolution based upon conflict man is also influenced by a higher rational mode of evolution based upon co-operation. The main principle of the latter is conservation of energy and control of man's environment and this is to be achieved by man's rational innovations in conjunction with the elimination of the wastes caused by conflict.

This concept of a higher rational mode of evolution building up by virtually refining and assimilating the lower conflict mode, is claimed as a significant conceptual innovation, and it is derived from the basis, outlined earlier in this work, of transferring the use of certain key scientific concepts from the more exact sciences to social science. This additional new concept of evolution is therefore, much more scientific and more rational than is the emotional appeal of Marx for bloody revolution and yet more conflict.

One cannot deny that Marx showed brilliant insight into the effects of energy harnessing innovations upon the whole social structure; that these effects were also, in some way inevitable, is difficult to deny. In that case, how can it be supposed that the bourgeoisie, who were the vital motivators of these rational innovations, could have been so uniformly ruthless and wicked as Marx seemed to claim they were? Might it not have been that some of them were consciously striving to achieve something for the good of mankind? Were there not influential people who tried to help the working classes at times when the rapid increase in population made poverty a very difficult problem to solve? These rising populations owed their very lives to the innovations in agriculture and industry. The grim Malthusian spectre of bare subsistence was a very real factor until advances in birth control stemmed the rapid rise in population.

To sum up, Marx appeared not to recognise or understand what I have described as the higher rational mode of action in man; he was careful not to attribute any good intentions to the bourgeoisie; and thus any action on their part was interpreted by Marx as contributing towards conflict. Since he appeared to see only the lower animal mode of conflict, Marx could only call for an intensification of conflict as a cure for the evils of his times. Finally, he assumed that after a period of savage bloodshed and social turmoil, quite automatically harmony and mutual regard would prevail. This does seem remarkably optimistic and indeed, unlikely as an aftermath to the horrors of bloody revolution!

By contrast the scientific approach of this book sees the process of change as a long drawn-out contest in which man is under the influence of a higher and a lower mode of evolution. In this contest there is no guarantee that the higher rational mode will prevail; but it is desirable that the higher mode should refine and absorb the lower; thus eliminating conflict and waste and promoting co-operation, conservation of man's energies and harmony. In this context the theory of Novicow is more relevant than that of Marx. Novicow argued that conflict should be refined into more intellectual and less physiological levels; this is in agreement with the view of evolution that I have tried to expound in this book; and also it points the way for human ideals and moral values.

16

THE HYPOTHESIS DISCUSSED IN RELATION TO THE THEORY OF WEBER

A large part of Marx's analysis was quite acceptable to Max Weber who saw his own work as a partial modification of the determinism of Marx's dialectical materialism. On the other hand Weber did not attempt to underestimate the very great influence of economic pressures. The theory expounded in this analysis is very much in agreement with both Marx and Weber with regard to the importance of economic influences.

My own view is that symbolic representation is one of man's most fundamental general methods or devices for harnessing and controlling resources and energy: and money is the general symbolic device that represents the value or the energy put in, in order to produce commodities; this is in accord with the labour theory of value as used by Marx; this point was made in a previous chapter and will be discussed in more depth later in this volume.

Weber emphasised the virtual inevitability of the increase in bureaucratic organisation. He held that the spread of bureaucratic form of organisation to all spheres is part of a general process of rationalisation in modern society. Weber regarded bureaucracy as a form of organisation that maximises efficiency in administration. Today many people, judging upon results would not agree; the subject is controversial, and Weber was unable to draw upon a scientific analysis of energy controlling devices; he, therefore fell into some ambiguities. The case that Weber was trying to make goes somewhat thus:

The use of all the most effective devices for harnessing resources and energy requires increasingly large commitment of resources, division of labour and specialisation; and these require also increasing size of organisation; hence the efficiency of the bureaucracy. This is, in theory perfectly true, but the fact that one has the most efficient devices for controlling resources, does not automatically ensure that they will be used in the way that is intended. Weber was very impressed by the rationality which could be imparted to an organisation by means of well defined rules and procedures; but these can fail in practice, since it is well known that one very quick way of stopping an organisation is by simply working to rule! There are always many customs, conventions or practices that have to be observed in order to make the rules themselves work. A great deal depends upon the spirit with which the rules are interpreted and carried out.

It has already been pointed out in this book that the distinction that Weber made between formal and substantive rationality is invalid; since one cannot separate the way of doing things from the objectives and the spirit in which they are done. Perhaps this is the reason why, above a certain size organisations become inefficient; that is to say, the individuals fail to identify with the objectives, lose sight of them or even pursue entirely other purposes than those of the company which they work for.

One point that I have made in criticism of Marx is very much in line with the views of Weber; that is that beliefs and ideas can be influential in social change. Weber's argument that the ethics of Protestantism paved the way for capitalism is very convincing.

In this present analysis a belief or idea may be considered as a device for harnessing energy and resources; group organisation, co-operation and mutual regard are, indeed some of the most potent methods or devices for conserving and controlling energy and resources. Thus concepts like limited liability and the Protestant ethic concerning capital accumulation, were as essential to the industrial revolution, as were the new machines.

17

THE EVIDENCE OF GORDON CHILDE

In his book *Social Evolution* Childe weighed the evidence for classification of man's history into ages or stages. He concluded that there were quite definitely no clearcut ages. Childe observes (p.31)

'The Maoris of New Zealand were still in a Stone Age when Captain Cook arrived in the eighteenth century A.D. The Stone Age in Egypt had ended before 3000B.C.! There is in fact no such thing as The Stone Age.'

Childe adopted the following general classification: Savagery which is assumed older than Barbarism; Barbarism which is assumed older than civilisation; and the highest stage in ethnographical evolution is civilisation. Childe took the revolution in food production, from cultivation of edible plants and the breeding of animals for food, as his criterion to define the passage Savagery to Barbarism: and he selected writing as 'a useful criterion' marking the transition from Barbarism to civilisation.

These are both key innovations leading to revolutionary developments in man's control of his environment by rational means. Childe states:

(p.33) 'I selected "food-production" as distinguishing the Neolithic from the earlier Palaeolithic and Mesolithic. Obviously the cultivation of edible plants, the breeding of animals for food, or the combination of both pursuits in mixed farming, did represent a revolutionary advance in human economy. It permitted a substantial expansion of population. It made possible and even necessary the production of a social surplus. It provided at least the germs of capital. And since animals and plants may be regarded as biochemical mechanisms, in cultivating them and breeding them men were for the first time controlling and utilizing sources of energy other than those provided by their own bodies.'

Childe quite explicitly identifies, as a key development, the energy harnessing and controlling innovations of food-production and cattle breeding. This is very clearly in agreement with the hypothesis presented in this book with regard to the significance of energy for social science.

Childe also observes: (p.33) 'As the technological criterion of the latest and highest of his "ethnical periods" Morgan took writing. I find it a very useful criterion. It may indeed seem odd that writing should be included in technology. But, after all, writing is a tool — an intellectual tool, if you will. It was the necessary instrument of exact science, the applications of which have revolutionised technology. Its use led to calendrical astronomy, predictive arithmetic and geometry — tools demonstrably used by the first civilised societies in the old and new worlds, by the Egyptians, the Sumerians, and the Mayas. At the same time a consideration of these earliest literate societies reveals that writing is a convenient and easily recognisable index of a quite revolutionary change in the scale of the community's size, economy, and social organisation.

'The invention of writing seems to coincide with a critical point in the progressive enlargement of the unit of cohabitation and in the accumulation of a social surplus.'

I see no reason to apologise for counting writing as a technological invention; it is a systematic organised process involving tools and it is a prerequisite to man's higher rational thought processes. Writing is, as Childe observes, an intellectual tool and its introduction is the motivation towards a rapid increase in man's rational thought and action. This has ultimately led to the great triumphs in technology that have harnessed enormous quantities of energy for present day industry.

It is demonstrated, by Childe, that although there are basic similarities in the various applications of these vital and powerfully influential innovations, nevertheless, the intervening steps in development do not exhibit even abstract parallelism. He discovered that there had been no parallel in developments in rural economy, weapons, cutting tools, use of metals, modes of transport, external trade, social institutions, warfare or the status of women. As an example of these divergences: (p.160) 'in Crete and temperate Europe as well as in Hither Asia wheeled vehicles were in use before civilisation was achieved, but on the Nile such were unknown for one thousand five hundred years after the foundation of civilisation. Here again we have divergence rather than parallelism, but in the Old World this divergence was rectified later by convergence; Egypt did in the end adopt the chariot.' This again demonstrates that there is no precisely deterministic time table or sequence of events.

In his conclusion, Childe points out that although there are significant similarities in the starting point of transition from savagery and the final pattern of the resulting civilisation, nevertheless there were both divergences and convergences in the intervening period. It is, therefore instructive to examine the concept of social evolution expounded by Childe in his conclusion:

Childe suggested that similarities do exist between organic and social evolution. (pp.162-3) 'organic evolution is never represented pictorially by a bundle of parallel lines, but by a tree with branches all up the trunk and each branch bristling with twigs . . . . In fact, differentiation — the splitting of large homogeneous cultures into a multitude of distinct local cultures — is a conspicuous feature in the archaeological record.

'But a comparison of the sequence summarised discloses not only divergence and differentiation, but also convergence and assimilation. To the latter phenomena it is hard to find an analogy in organic evolution.

'Total replacement of one society or culture by another is not the typical form of convergence and not that generally observed to lead to civilisation. Two cultures may generally become more alike without losing their distinctive individualities. The same novelty may appear simultaneously in two different cultures, or first in one and then in another; both consequently becoming more alike.'

Examples of cultural borrowing or diffusion between politically and culturally distinct societies are: the adoption of railways by Russia and Japan from Britain; and the adoption of war chariots by Egypt and Greece similar to those used in Mesopotamia a thousand years earlier. In none of these cases could conquest of one by another be inferred.

Childe points out that (p.171) 'with certain modification the Darwinian formula of "variation, heredity, adaptation and selection" can be transferred from organic to social evolution, and is even more intelligible in the latter domain than in the former.'

The process of cultural variation is obviously more rational than its counterpart, biological mutation, since the former depends upon intelligent action while the latter occur by accident. The mechanism of social heredity is much more rapid than the biological mechanism of sexual reproduction since it is effected by example and precept, by education, advertisement and propaganda. These are more intelligible and rational processes than are those of biological heredity. Once a set of inventions and devices for controlling a community's environmental resources have been acquired, they can be quickly passed on to the next generation. By the process of biological mutation a particular change can take hundreds of generations.

Adaptation to the environment is as much a condition of survival for societies as for organisms. In social or cultural adaptation, the process can be much faster than in biological organisms because it can be taught and transmitted between people; again adaptation, in the social sphere, can partake of a much higher degree of rationality than can biological adaptation.

In the process of selection, once again rationality is much more in evidence in the social than in the biological sphere. In the social sphere it is often not the individuals that struggle for survival, but it is the ideas or innovations that are selected or rejected; in turn, the survival of the men and of their communities depends upon the judicious selection of innovations that will be effective in harnessing the energies and resources of his environment to serve man.

Overall, it is the greater potential for rationality that distinguishes the human social evolution from the animal biological mode of evolution; however the Darwinian formula of 'variation, heredity, adaptation and selection' applies to both.

The rational content in social evolution manifests itself in the diffusion of innovations which enable men to control and harness the energies and resources of their environment. This in turn accounts for the phenomenon of convergence into civilised forms of very similar kinds. There is a logic of choices and development, which, if pursued, leads towards similar conclusions. However, the phenomenon of divergence can be accounted for by the facts that men may reason differently, they may have differing amounts of knowledge, different calibres of will to pursue their purposes, and differing available resources.

The logic of energy processes certainly acts as a powerful constraint guiding the development of man in society, along certain paths; however, there is nothing of a deterministic nature in these constraints, since they operate through the medium of man's ability to choose; and there are societies that do not follow such logical paths; they may be intellectually incapable, or they may prefer the less affluent consequences of the alternative courses which they have chosen; possibly some communities are held in some form of subjection by more powerful nations. There are today, the underdeveloped nations which may one day follow the same patterns of development as the more advanced; however, neither the timetable nor the course of such advances are predetermined; and some may never progress in this way.

The foregoing is a résumé of the concepts of evolution outlined by Childe in his conclusions, interpreted in the light of the energy analysis which is expounded in this book. Childe puts forward a very convincing set of ideas on social evolution; it is all the more valuable because it is firmly grounded upon archaeological and historical evidence. It would be misleading to attribute the general remarks about energy in the foregoing account to Childe, although he did make one statement in the energy context; and this is quoted in the preceding discussion.

The addition of the energy analysis to the evolutionary concepts of Childe imparts to them a unifying theme and a cohesion which they do not otherwise possess. Despite the care and detailed research with which Childe argues his theories of social evolution they do not fit together as one, without the complementary principles of rationality and conservation of energy to knit them together. Childe does mention energy and also he repeatedly refers to social evolution as more intelligible than organic evolution; however it is necessary to be rather more emphatic and to give more explanatory detail of the way that the principle of C.O.E. influences social evolution. The analysis of this book is intended to fulfil this requirement. This volume also emphasises another theme that barely emerges in Childe's account, namely that man partakes of both the higher rational mode and also the organic or lower animal mode of evolution. Nevertheless Childe's account is of great value.

18

THE VIEWS OF ROBERT REDFIELD

In his book *The Primitive World and its Transformations*, Robert Redfield discusses the relationship between what he describes as the 'technical order' and the 'moral order'. Briefly, his views are: (p.36) 'In folk societies the moral order predominates over the technical order. It is not possible, however, simply to reverse this statement and declare that in civilisations the technical order predominates over the moral. In civilisations the technical order certainly becomes great. But we cannot truthfully say that in civilisation the moral order becomes small. There are ways in civilisation in which the moral order takes on new greatness. In civilisation the relations between the two orders are varying and complex.

'The great transformations of humanity are only in part reported in terms of the revolutions in technology with resulting increases in the numbers of people living together. There have also occurred changes in the thinking and valuing of men which may also be called radical and indeed revolutionary innovations. Like changes in the technical order, these changes in the intellectual and moral habits of men become themselves generative of far reaching changes in the nature of human living.'

Redfield identifies the transformations from primitive times to the modern era with a continuous process of breaking down and rebuilding of the moral order:

(p.58) 'As the technical order develops with the food producing and urban revolutions, as the civilisations produce within themselves a differentiation of human types, and as they also reach out to affect distant peoples, there is a double tendency within the moral order. On the one hand, the old moral orders are shaken, perhaps destroyed. On the other hand, there is a rebuilding of the moral orders on new levels.'

However, Redfield does not consider that the moral order is simply acted upon by the technical order; and he expresses the view that (p.83) 'ideas, generated early in the course of technological development, became themselves causative agents of further transformation in human living' and (p.84) 'With the development of writing, literate and reflective people, and enlarged opportunities to travel, to communicate and to think things over, the power of ideas to create ideas and of ideas to create institutions greatly increased. Some of these ideas — some of the powerful ones — have to do with the right, the good, and the true. We may describe this change by saying that from now on the moral order is self-regenerative.'

Redfield states (p.89) 'Those ideas in history which have the most force are those which speak for everyone.' Redfield cites Whitehead in this context: (p.90) 'He makes us see that such an idea has power of development that is recognisable in and yet apart from the particular occasions on which it is enunciated . . . . says Whitehead the idea of human dignity abolishes slavery and goes on to demand that there be no more second-class citizens, that forced labour of the innocent, and that the indignity of racial segregation come to an end.

'So following Whitehead's lead, we may suspect that other ideas of corresponding power and endurance are already at work among us: the idea of permanent peace, also the idea of universal human responsibility, to balance and extend the creative idea of universal human rights . . . they are unrealistic, fanciful, Utopian. So they are; but also they are among the movers and shakers of human affairs. And their strength lies in their universality.

' . . . . Only civilisation could bring about the circumstances of moral conflict in which these ideas could arise and the means for their transmission and reflective development. Civilisation is a new dimension of human experience. The great idea moving among many traditions and in newly troubled minds, is now an agent of change, a shaper of the moral order.'

In his final chapter, Redfield concludes:

(p.165) 'The moral canon tends to mature. The change is far from steady, and the future course of ethical judgement, is not, it seems to me, assured to us. But in this sense — that on the whole, the human race has come to develop a more decent and humane measure — there has been a transformation of ethical judgement which makes us look at non-civilized peoples, not as equals, but as people on a different level of human experience'.

Redfield may be right; of course we do not abandon our dying kinsmen as do the Siriono of the Bolivian forest nor do we expose unwanted infants to die; however, there are still the ethical problems of abortion, modern warfare and possibly euthanasia to be accounted for; not everyone would see these as moral advancement. They might be justly described as moral maturing since it appears that the motivations for the modern problems is very similar to those of their more primitive counterparts; euthanasia is not in principle very different from the abandonment of dying kinsmen in the forest; abortion is similar in principle to exposing unwanted infants and modern warfare is even more destructive than the fighting of primitive men; in this activity the morals of man have matured so far that he now threatens to destroy and render the whole world uninhabitable.

The work of Redfield is of very great value in the field of anthropology; but he has expressed various doubts concerning the moral order. In particular, he concludes that it is difficult, if not impossible to remain objective in making ethical judgements. This must be true since it is the individual who makes the judgements; and his knowledge and rational capacity is limited; he cannot stand apart and see every facet of humanity without involvement. His judgements and decisions affect his own future; he is part of the object under observation. A man's moral theories or beliefs constitute a model of the way *he* sees the moral order.

It is rather surprising that having cast considerable doubt upon the ethical neutrality that is held essential to the standpoint of 'cultural relativism', Redfield then criticises Ruth Benedict on the ground that she, a cultural relativist, fails to maintain ethical neutrality. My own opinion here is that there is much to be said for neutrality when we are collecting data scientifically and when we are testing it; however when we are trying to put together a theory, it is difficult and often unwise to ignore ideas and concepts that have been previously advanced. In my view, even the most unbiased scientist will assert what they believe to be true, even before full evidence is made available. That is what Ruth Benedict appeared to be doing in the passage criticised by Redfield.

Redfield observes: (p.153) 'Ruth Benedict was a cultural relativist who told us that cultures are equally valid. Nevertheless, in reading some of her pages, one doubts that she found them equally good. In the seventh chapter of "Patterns of Culture" she introduces the concept of "social waste". Here she leads the reader to see a resemblance between the values of Kwakiutl society and those of his own (Middletown); both emphasise rivalry. But rivalry, wrote Benedict, is notoriously wasteful. It ranks low in the scale of human values. One asks whose scale? Is there a universal scale of values which ranks rivalry low? She goes on to point out not only that "Kwakiutl rivalry produces a waste of material goods" but also that "the social waste is obvious" . . . . The line between description and evaluation is here unclear. It is very hard to say that culture A produces more suffering and frustration than does culture B without saying also that in this respect you prefer culture B.'

In my view, Ruth Benedict was most certainly not maintaining ethical neutrality; she was in fact producing one of those great ideas to which Redfield himself refers: (p.91) 'The great idea moving among many traditions and in newly troubled minds, is now an agent of change, a shaper of the moral order.'

Redfield himself, perpetrates a similar departure from ethical neutrality when he writes: (p.143) 'Writing of Petalesharoo, the Pawnee Indian who in the face of the customs of his tribe rescued a woman prisoner about to be put to death ceremonially and strove to end human sacrifice among his people. I called him "a hint of human goodness". Plainly I placed a value on his conduct.' Petalesharoo had no ethical neutrality to guide his lonely choice, but few would deny that he was right to deplore the social waste and suffering caused by human sacrifice; for there can be no greater waste than that of life itself. Redfield also departed from ethical neutrality to approve of what Petalesharoo did. Ruth Benedict also departed from ethical neutrality in deploring social waste. I believe that this also represents a concept of great significance and, if I may borrow Redfield's phrase 'a hint of human goodness'; for to reduce the wastage due to conflict and replace it with the economies of co-operation must be good. I have arrived at the same conclusion as did Ruth Benedict by a series of scientific deductions but not, I think by ethical neutrality.

19

MORALITY DEFINED

Morality must be defined in a correct sequence: thus, right action must be preceded by the right attitude or spirit of co-operation and mutual regard. This will lead towards right and rational thought; which, in turn, leads towards right and rational action. By reducing conflict this reduces social waste and promotes the economies of co-operation.

The above paragraph defines morality in the context of C.O.E. The wastages which are to be particularly avoided are those mentioned in the previous chapter on rationality, namely, profligate waste of resources, pollution of land, water, and air, and destruction by warfare.

The correct sequence is vitally important; any attempt to start with the aim of conservation of resources and work through towards the right attitude is liable to lead to selfishness and conflict. This is the course which men have most frequently tried to take and it has always ended in disaster.

I cannot claim to have remained ethically neutral since the source of my inspiration has been as follows:

(Matthew 6:30-33) 'Wherefore, if God so clothe the grass of the field, which today is, and tomorrow is cast into the oven, shall he not much more clothe you, O ye of little faith? Therefore take no thought saying, What shall we eat? or, What shall we drink? or, Wherewithal shall we be clothed? But seek ye first the kingdom of God, and his righteousness; and all these things shall be added unto you.'

I take this to mean that if we get our priorities in the right order, then our problems of scarcity of resources will be solved by the reduction in social waste and the proliferation of the economies from co-operation.

It is important to further inquire, what briefly and precisely is meant by 'the kingdom of God'. Here we must again consult the Gospel (Matthew 22:35-40) 'Then one of them, which was a lawyer, asked him a question, tempting him, and saying, Master which is the great commandment in the law? Jesus said unto him, Thou shalt love the Lord thy God with all thy heart, and with all thy soul, and with all thy mind. This is the first and great commandment. And the second is like unto it, Thou shalt love thy neighbour as thyself. On these two commandments hang all the law and the prophets.'

The first of these great Commandments, I take to embrace the enlightened search for Truth (God), learning, education and science for the sake of the higher refinement of man's activities.

The second of these great Commandments, I take to embrace the mutual regard essential to man's co-operation in order to control his environment and conserve his resources. It should be emphasised, however, that the primary aim should be the appreciation and the belonging of human kind in society; the resulting material gain will be in the nature of a well-earned by-product. That is what I see as the meaning of the foregoing quotations from the Gospels. If we look for the material gains first, then we usually prize this above all else and we will never achieve the primary objective which should be mutual caring and belonging.

Just as Redfield sees 'a hint of human goodness' in the enlightened actions of Petalesharoo, so also more than a hint of human goodness in the life and principles of Jesus, can be the unerring guide in assessing all moral developments, whether they be improvements or deteriorations. The principle of conserving and controlling resources is very important but it needs to be kept to its proper place and function in our order of priorities. Mutual regard and caring must come first.

The moral order should always take precedence over the technological order. It is the guiding spirit, without which conflict, disruption and disaster will engulf human kind.

It is also clear that rationality and goodness are, when they are defined with strict logic, identical; and the moral order can be identified with 'the higher rational mode of Human behaviour' as defined in the previous chapters of this book.

20

DISCUSSION: DEVICES FOR HARNESSING AND CONSERVING ENERGY

Since there is nothing deterministic in the way and timing of innovations, it is not possible to give a strictly chronological historical sequence for the development and introduction of inventions. It is, however, necessary to draw attention to the range of different kinds of devices which can be involved in harnessing and conserving energy.

The most obvious energy harnessing devices are prime-movers like steam engines. Tools and other implements, such as the axe, spear, needle and potter's wheel can easily be seen as means of applying energy in a controlled way in order to achieve desired effects.

It is a little more difficult to see various methods of symbolic representation as energy harnessing devices; however, a little reflection reveals how indispensible are devices like speech, writing, arithmetic and mathematics to the effective use of modern machines and tools; they were also essential to the earliest civilisations of Egypt and Sumeria.

Similarly, various means of communication are also essential to the proper functioning of modern means of energy in productive processes; these include telegraph, telephone, printing, and tape control.

Various methods of group organisation are just as necessary to modern industry as are sophisticated machinery. Therefore devices such as limited liability, government, and cost control; as well as concepts like authority and obedience are also essential to modern methods of controlling energy for industry. These are also supported by devices like rules, regulations, accounting methods and scientific laws.

A further indirect influence is exerted upon social activities by feelings, beliefs, attitudes and ideals. This can be illustrated by the influence of the Protestant ethic upon industry and commerce and also influence of *laisséz-faire* on all aspects of social organisation in the early nineteenth century. In the same general category are to be placed the attitudes of mutual regard or alternatively of distrust which can promote either, on the one hand, economies from co-operation or, on the other hand, the wastes of conflict.

It may be that influences of feelings, attitudes or ideals appear, at first sight to be insignificant; but further consideration reveals that co-operation and conflict are the greatest influences either for economy in social resources or for waste, according to whether the feelings are of mutual regard or of distrust. It is in these attitudes that the purpose to which the energy controlling machines and devices is decided, whether constructive and good or wasteful and evil. Thus mutual regard and co-operation rank as the most powerful of all devices for the harnessing and conserving of energy and resources.

For example, wheeled transport can be used for productive work or for war chariots. Warfare may stimulate some noble instincts and inventions; but, for the duration of hostilities the scope of human co-operation and regard is drastically curtailed; thus energy controlling devices are perverted to the very worst of immoral purposes and the most grievous of social wastes; namely, the destruction of human life itself. The more sophisticated the energy harnessing devices are, the greater is the waste and destruction. If the spirit were right; why should we not have all the new inventions without the waste and destruction? It is surely a matter of putting the resources to the right use in research rather than the wrong use in warfare.

Only a few examples of each type of device have been given; listing and describing the whole range of devices required for harnessing energy would occupy a whole volume; but that is not the purpose of this book. All that is possible here is to illustrate the fact that there are, as well as machines, other devices such as tools, means of communication, methods of symbolic representation, rules, regulations, methods, theories, group organisation, feelings, beliefs, attitudes and ideals which all contribute to the total social situation in which the machines are made to work; and they all exert influence upon the overall results.

However, there is one device which plays an extremely vital role in the transition of man to civilisation, namely the symbolic device of money; again, it is not intended to undertake any historical treatment of the subject, but it is the conceptual transformation of man's relationships that is significant, rather than any historical sequence. Money is an idea which is a powerful formative influence; and this influence is largely instrumental in creating a situation in which the higher rational mode of thought and action could develop.

At the time when man begins to gain greater control over his food supplies, the security of his life is made safer; this is only a tendency; since complete security of life is not ensured even today when starvation faces millions every year in the under-developed countries. However, in those areas where food has been made abundant, this improvement has been achieved by means of modern machinery with division of labour in a monetary economy.

In such a situation when life is no longer precarious, men can choose those ideas, devices and innovations which contribute most to their satisfaction of wants. In this way the struggle for survival can be transferred from the individual organisms to the ideas; since only the most effective ideas or inventions are likely to survive: thus, a significant step forwards from the lower animal towards the higher rational mode of evolution can be achieved.

21

THE ENERGY BASIS OF ECONOMICS AND MONEY

The transference of the struggle for survival from individual organisms to the innovations, operates in a monetary economy by a method analogous to the death of an organism. The way in which this operates is to be found in modern economic analysis, where it is postulated that, in order to avoid making a loss when an additional article is produced, the marginal cost must not exceed the marginal revenue; that is, the extra cost incurred by producing an extra article must not exceed the addition to the total revenue so obtained. In economic theory, it is argued that if loss is incurred the company will either become insolvent or change to a different line of production.

It is the contention of the theory presented in this book, that, in a monetary economy insolvency acts to eliminate the identity of any group or company that consistently makes a loss; and an innovation that is not considered to effectively satisfy human wants cannot be sold; and therefore, will be eliminated. Thus, the less effective ideas are eliminated while the more effective ones are retained in use.

The basic assumptions of economics are that the resources controlled by man are limited while the wants of man are unlimited. This fundamental limitation of man's resources arises from the limitation of the energies in any system. All matter is built up from energy; and similarly all the commodities of men are constructed by the energies or labour of men upon the energies in matter.

This fundamental factor of energy was recognised by Karl Marx in the labour theory of value which states that: The value of any commodity is determined by the quantity of average socially necessary labour required to produce it. However, the energy in question need not be solely of human bodily source but may also be from other sources of power harnessed by means of machines under man's control. Nevertheless, every assessment of value that is made is a personal judgement and such judgements rarely involve any very precise energy calculations: money is, therefore, only a very rough symbolic representation of the inherent energy aspect of man's activity. Nevertheless, it is a vital factor which must be emphasised, that by a process of development, and natural selection, money has been evolved which man uses, albeit inadvertently, as an approximate measure of value and of energy put in to production of commodities. Of course, it can be argued that price is determined by supply and demand; and, therefore is not connected with the energy or labour put into production. However, in Britain, there has been a good deal of legislation concerning prices and incomes; which has been specifically designed to establish some sort of relationship between prices, incomes and the level of production. Moreover, much of this has taken place under the influence of socialist theories: thus the labour theory of value has been undoubtedly influential. These have been attempts at rational control of the economic system; they have only had limited success; but the point that I am making is not intended as political. It cannot be denied that some efforts have been made to implement the labour theory of value; and by my previous reasoning this implies a rough measure of socially necessary labour, or other energy put in.

It is also worthy of note that most of the energy used in production is derived not from human muscle power, but from vastly larger resources of natural energy by means of machines. The energy analysis is not simply an idea originating in the recesses of my brain; on the contrary the energy processes are out there for all to see quite clearly. As a measure of value energy can be no more accurate than the rationality that has gone into the logical devising of the economic plans; this, I feel, can be fairly described as rough in the present context. Therefore, the above claim that money is a rough measure of energy put into production, is vindicated.

The device of money as a symbolic representation fulfils a key role in man's development. In a very remarkable way, the symbol of money and the thing that it represents tend to become identified together in a very intimate relationship; so that the monetary symbol and the actual valued article are rapidly interchangeable. It has often been said that paper money is of no value; nevertheless, people do treat it as though it were valuable, because it can be quickly exchanged for valuable goods. A person with a lot of money, even in the form of bank balances is regarded as being of great wealth and substance.

By implication, therefore, money is equivalent to the labour or the energy that the individual would have to expend if he could make the goods himself. It is the facility for rapid exchange between the symbol of money and valuable goods, which gives the simulation, in social life, of a system of rapid energy transformations; thus, instead of having to make the goods himself, by relatively slow energy processes, or obtain them by lengthy bartering, the individual can put his own energies into that which he does best in exchange for monetary tokens which are agreed to be the rough equivalent of his labour; this money he can exchange for the products of the labour and energies of other people.

It is not suggested that this process is perfectly rational or absolutely fair; on the contrary, it is usually very rough and often unfair. If we can leave aside his insistence upon bloody revolution and conflict, there is still a great deal of logic in the arguments of Karl Marx. The important fact is that there is an energy process taking place and that in this process money plays the role of a symbol which represents energy in a rough and ready fashion. If man's rationality and planning were better, then the processes could be less rough and more just.

Therefore, money is the outstanding example of a symbol that 'becomes' the thing that it represents and its identity merges with both the value of the commodities which it buys and also the energy or labour expended in their production.

The whole of the economic system is, in some sense, a tribute to the unique rationality of man, that he is thus able to similate the workings of the laws of energy, in some measure and thereby obtain the economies in energy and resources which arise from co-operation and division of labour. He has understood, to some extent, the nature of the principles that constrain the materials of his body and of his environment; and he has devised economic systems using the symbolism of money to influence his activity accordingly. The processes are not perfect but they are vastly more rational than the efforts of any other species.

Much more might be said upon this subject; but this book is only a very rapid general overview of man in society. Therefore the matter must be left there for the moment.

22

THE SOCIAL SYSTEM

At this point in the present analysis, all of the ingredients of the intended scientific model of man in society, have been discussed; it now remains to put these together to form the overall model.

Systems analysis is a general rational method which can be applied to a great variety of different systems. It has been made clear in this work that a deterministic model is quite inapplicable to man in society. In fact it has been suggested in the previous chapter on rationality that man is at the apex of a hierarchy of systems, built one upon the other ascending in order of rationality of their controlling networks. Each higher system builds itself by using the lower systems as their bricks or raw materials.

The lowest is the inert which is purely mechanistic and apparently lacking in rationality; the next higher kind of system is the living organism; some of these possess a sort of instinctive rationality of a low order; the next higher system in the hierarchy is the individual man with a generally much higher rational potential; and the highest possible potential in rationality can only be envisaged in the system at the very pinnacle of the evolutionary hierarchy, namely 'man in society'. Throughout each of these systems the law of C.O.E. are influential as a constraint; and an energy balance is a normal part of all systems analyses.

The human social system depends upon some freedom of choice within the constraints of C.O.E. The components of a deterministic, physical system appear to fulfil their purposes by contributing to the overall functioning of the larger system in which they are involved: but individual men have such a great variety of objectives, that the overall social system must somehow manage to function without imposing a mechanistic regime or ritual upon its members.

This is one of the great problems of modern man, namely, that efficiency is often sought in industry and commerce by the imposition of machine-like rules, regulations and rituals. This often fails because of capacity of individuals to introduce variety in the shape of their own private objectives; and their ability to defeat the purpose of the rules. Industry and commerce are not, however, the whole of man's life; there are, in industrial countries, a great variety of skills and occupations and usually a good deal of leisure, recreation and activities other than work. Unfortunately, there are also in the less developed nations, social systems which produce little or no surplus for recreation and in which life itself is extremely precarious.

The views of Durkheim are relevant in this context; he considered that the 'organic solidarity' necessary to bind together a modern, complex society, was provided by the social division of labour, which thus had a moral character. Durkheim did not consider the mutual benefits of exchange to be an adequate basis for social integration; but he held that mutual dependence involved the inseparableness of one's image from the images of those who are intimately interdependent with and complementary to us. This, in my opinion, corresponds with my own view that morality is associated with the mutual regard that should motivate rational co-operation; whereas the benefits of exchange are a desirable byproduct of this fundamentally creative attitude.

'For where interest is the only ruling force each individual finds himself in a state of war with every other since nothing comes to mollify the egos, and any truce in this eternal antagonism would not be of long duration.' (Durkheim 1964 pp.203-4)

Durkheim contrasts this organic solidarity with what he describes as the 'mechanical solidarity' in the simpler structure of feudal society prior to industrialisation when resemblance of the individuals rather than differentiation, was the outstanding feature. Thus the change from feudal to industrial society is dependent upon the increasing use of man's potential for rationality, division of labour and the economies of co-operation.

It must be reiterated, however, that the lower animal mode of behaviour still pervades the more rational modern scientific behaviour thus perverting it to forms of conflict whereby the gains are largely dissipated; so that rational and moral improvement do not achieve their fullest potential.

In any investigation into the use of man's energies, it is very easy to worry about the apparent wastage of energy due to the great variety of the activities of the individuals in making use of their freedom to choose. Such worries are quite unfounded since they arise from the pursuit of an illusion of a completely unworkable deterministic human system. In this connection the important questions that need to be resolved are concerning the nature and relationship between the objectives of the individuals and of the overall social system; the vital questions are; What does the individual do for the overall system? And what does the overall system do for the individual? Ideally, the individual must act in co-operation and in mutual regard for his fellows so that each cares for and feels wanted by the others. This, under modern rational scientific conditions provides a substantial surplus which can be used for leisure and recreation by the individuals. The overall social system with its various authorities and controls, have the duty of providing the right formal structure in which these activities can take place; resources are, of course, provided for the work of the authorities by the individuals.

Society has no objectives or functions of its own other than providing the best environment for the individuals. Therefore, the leisure, recreation and other personal objectives, which might appear on deterministic assumptions to be losses and wastes from the system, are in fact the vital ends and objectives of the system.

It is unfortunate that the rationality of man frequently fails to attain its higher potential; because of the waste of resources due to selfishness, conflict and the lower animal mode of behaviour. This duality in the human social system must be represented in the detail of the overall scientific model.

It remains to consider what type of system is appropriate to the kind of social system described above. The concept of homeostasis or dynamic self-maintenance does not allow for the fact that society may change its form rapidly in order to utilise new inventions or methods of using energy. Parsons seemed to neglect this fact and assumed an overall stability which is of very limited application to man in society.

The concept of morphogenesis postulated by Walter Buckley in his article 'Society as a Complex Adaptive System' is much more appropriate to man in society because it recognises not only the dynamic self-maintenance of the system but also its capacity for changing its form and structure.

It has been demonstrated in the previous pages of this work that the principle of C.O.E. influences man's freedom of choice, his rationality and also his morality. The concept of C.O.E. has also been shown to be complementary with the principle of uncertainty. Both of these concepts are relevant to systems methods as also are the concepts of entropy, negative entropy, feedback and equifinality.

The principle of equifinality states that final results can be obtained from different initial conditions and in different ways; thus the social system is not restrained by the simple cause and effect relationship of closed systems. Closed systems are subject to the influences of entropy which increases until eventually the entire system stops; this movement towards maximum entropy is also a movement towards disorder. In a closed system the change in entropy must always be positive; however, in the open, biological or social systems, entropy can be arrested and may even be transformed to negative entropy, a process which approaches more complete organisation.

The only way in which the open system can offset the tendency towards increasing entropy is by continually importing material, energy and information in various forms; and then transforming and redistributing these resources. Some are incorporated into various parts of the system and other byproducts are returned to the environment. These processes of dynamic self-maintenance can only be achieved by the use of an adequate controlling system which can vary in complexity from the very simplest organism through the instinctive to the highly complex human brain.

Human rationality is essential to the social system in order that, knowing the underlying energy processes, men can make the logical dispositions of the system's resources; so as to satisfy the individual objectives and also maintain the dynamic self-maintenance of the whole system. However, we must always remember that rationality is never perfect since man has a lower as well as a higher mode of behaviour.

The principle of uncertainty is also applicable because, although the principles of C.O.E. are the general basis of rationality, nevertheless, they do not act deterministically. It is by rational thought and action that men can co-operate and so conserve their scarce resources. This entails variability of behaviour, or what has been described as 'free will' within constraints. It is this variability of choice that enables man to have moral aspirations.

The principle of entropy is also relevant to morality, since it relates to the process of taking in resources and energy and using them to provide dynamic self-maintenance for the individuals and the social grouping. This must be based upon conservation of resources and in particular upon the preservation of the most valuable resource of all, which is human life itself. Morality is founded, therefore on respect for human life, mutual regard, co-operation and the resulting communal economies and benefits.

There I must conclude this very brief account of systems as applied to man in society. It must be again emphasised that the whole theme of this book concerns scientific sociological theory; and none of the theory used is in any way deterministic. Therefore the systems methods proposed here are not in the nature of computer programming techniques, since these are in no way applicable to the system of man in society. However, systems thinking is used in order to elucidate the overall conceptual form of a realistic model of man in society.

23

THE CO-ORDINATION OF THE VARIOUS SOCIAL THEORIES

In bringing this short work towards its conclusion, it is appropriate to consider what contribution it can make in elucidating the problems and ambiguities that have arisen in social theory. Sociologists have levelled many criticisms against their own discipline; and these have shown up as conflict between rival 'schools'.

This conflict in sociology mirrors the continuing conflict in man's affairs; and its modern form is expressed in the paradoxical tension between 'social action' and 'social systems'. These are overall labels which cover numerous viewpoints or perspectives in sociology; but broadly they pose the question of whether man creates society or alternatively, society creates man.

This dichotomy of thought is a symptom of the modern fragmentation of knowledge; which derives from the enormous growth in man's knowledge and the specialisation of disciplines necessary to cope with it. In earlier times it had been clearly understood that knowledge was unified; and this was enshrined in the ancient principle of 'Unity in Truth'. It was also clear that man and society were intimately interwoven; so that, as Aristotle put it, 'one man is no man at all'; interdependence among human individuals is not merely a convenient arrangement; but is part of the essential character of human nature.

These ideas present a more integrated view of man in society than does the argument between social action versus social systems. Nevertheless, the older ideas were based on mere scraps of knowledge compared with the vast quantities of data available today. Perhaps the ancients owed their unity in thought to the facility for generalisation which exists when detailed knowledge is scanty; and also to the close-knit nature of their social life.

It is certain that science has provided masses of detailed data and has nurtured the specialisms that have so greatly proliferated and fragmented man's knowledge and also his social life. However, it is the thesis of this book that science, by way of compensation, also provides in C.O.E. the unifying concept whereby the fragmented disciplines may be joined and interrelated.

The crude ambiguity of the systems versus action controversy can be expressed in the question of whether man controls society or society controls man. The bald oversimplicity of these crude questions highlights their complete inadequacy for the analysis of complex social and individual relationships. The word control is simply not relevant in the sweepingly deterministic sense which is implied in ordinary linguistic forms. The intimately interrelated nature of the bond between man and society is a much more subtle and variable arrangement than one is able to convey by simple concepts like control; the influence exerted by means of energy in the context of free choice is much more varied and open to rational refinement than the rigid formula of control would suggest. For this reason the scientific energy analysis is capable of providing the vital unifying relationships.

The refinements of suggestion which are used by the modern media and by large organisations can be very effective; but they need not fall under the rigid category of control.

The idea of self-control based on freedom to choose in the context of constraints indicates some degree of influence in both directions, that is, man upon society, and society on man. These also imply moral choice and rationality, neither of which could be very meaningful if control were unilateral and rigid.

Therefore, the question arises as to whether action and systems aspects are complementary or contradictory. This issue is rendered more acute by the close intermingling of the two quite contrasting modes of human action, namely the lower animal and the higher rational. The former involves a great deal of conflict and is conducive to rigid domination or control; whereas, the latter requires a high degree of co-operation and a spirit of mutual regard of an entirely higher order.

In practice, a good deal of moral and righteous indignation tends to enter into the argument between social action and social systems. This issue cannot be resolved simply by considering the 'facts' as they exist; since regard for the individual and for society or opinions about conflict, are vital aspects of what the protagonists feel should be made to happen.

The scientific analysis of this book, enables a more objective appraisal of such matters; here moral behaviour is considered generally to involve mutual regard and co-operation which reduces conflict and conserves man's resources. Moral behaviour is, in the final analysis, ideally rational with the individual freely choosing the paths of co-operation and building order rather than conflict and destruction. That these arguments are carried out using the scientific concepts of C.O.E., indeterminacy, entropy and systems, should reduce the all too familiar emotional and irrational content.

There are, of course polluted forms of rationality which extoll the positive virtues and benefits of conflict and even of warfare. The philosophies of Hegel and Marx are examples of such perversions of rationality. These are the rationalities of man dominating by warfare, violence and bloodshed; and they have culminated in the Nazi slaughter of the Jews and the Siberian forced labour camps. It is clearly the divisive fallacies of these polluted ways of thought that perpetuate conflict, war and destruction of human substance.

The inherent dichotomy and tension between man's two modes of action are reflected in similar conflicts and ambiguities in the divisions between the major contenders in the arguments between sociological theories:

Thus Cohen refers to the atomistic and holistic approaches where the former tends to treat social wholes as possessing characteristics similar to mechanical objects; that is, composed of identical replaceable parts which can be assembled in different ways; this Cohen described as the action theory of society. The incompatibility between the mechanical analogy and real regard for the individuals is abundantly clear.

Cohen refers to the 'holistic' as treating societies as having characteristics similar to those of organic matter. This approach has the advantage of avoiding the mechanical difficulties; but the organic analogies, though suggestive are too general and imprecise to describe adequately the unique features of man in society. The systems approach needs to be more precise in the building of its model in the way proposed in the previous chapter.

Cohen considers that the duality displays two complementary aspects which are reconcilable. By contrast Dawe regards the two aspects of man as fundamentally opposed conceptions of human nature and the way man relates to society. Dawe disagrees with Nisbet's view that sociology developed through the nineteenth century concern with the problem of order, and reaction against the individualism of the enlightenment, and its expression in the French and Industrial Revolutions.

Dawe argues that the problems of order and control constitute the source of two distinct and conflicting sociologies. A sociology of social system with the notion of the actor as being controlled by the system through socialisation and social sanctions: and alternatively the sociology of social action with the actor defining the situation and acting upon it.

The first of these two views was expressed also by Thomas Hobbes, in almost mythical form; he postulated an original state of nature in which man's life would have been 'poor, nasty, brutish and short'. This pessimistic view of human nature, Hobbes considered, could only be avoided by society constraining the individual by means of authority, socialisation and sanctions.

The second of these views appears to correspond with the optimistic Enlightenment view of man which held that man should create and control his own social situation and that it is unnatural for him to be in a state of subjection to the social forms.

It is quite easy to recognise in these two distinct and conflicting views, the pessimistic Hobbesian and optimistic Enlightenment views as being identical with the two categories referred to in the previous pages of this analysis, namely, the lower animal, and the higher rational mode of behaviour. Hobbes is obviously basing his work on a backwards look at a postulated original, primitive, animal state of man; while the Enlightenment view looks forwards to an ideally rational state of man. The contention of this book is that neither of these states of man exists in pure or isolated form; but they both exist together, intimately interwoven in man's social life.

I am aware that some people suggest that primitive men and many animals were essentially peaceful; however, it has been observed that nature is red in tooth and claw; thus, if the basis of animal evolution is to be accepted as struggle for survival, then aggression, at least towards other species, must characterise those animals which survive. It is a short step from primitive man killing other animals in self defence, to killing them for food; and since men are unlikely to have starved to death peacefully, it would have required a period of super-abundance of food growing wild, to support a situation in which men are likely to have remained at peace with each other. Of course such super-abundance may have arisen and may also have corresponded with the mutation which gave rise to human species. If these speculations were correct, then it would lend certain 'Garden of Eden' qualities to man's origins which many would be rather loath to believe in; however, this would support rather than invalidate the basic assumptions of this analysis; since any end to the super-abundance would present man with a very serious moral problem, namely how to survive without killing his fellow men who were competing for the scarce food.

There is plenty of archaeological evidence of cannibalism, violence and murder by primitive man. Therefore, it appears that any peaceful phase ended quite early in man's evolution; and gave way to a period in which man reverted to much of his previous lower animal behaviour. However, the germ of rationality having been set, men would henceforth, always have some higher rational content intermingled with the lower animal aspects of their activity. This would still remain true even if the new species 'rational man' were to interbreed with prehominoids not of the same mutation. To put it bluntly we could have ended the evolutionary trail as humans with a rather mixed collection of motivations. I do not apologise for describing the violence, murder and selfishness of man as lower animal behaviour; for these were, in my view, characteristics of the prehominoid animals that gave rise to the mutation of man.

The Hobbesian and Enlightenment views of man are based upon myth and speculation; but as is often the case with myths there is a sound core of facts upon which these myths have grown. It is interesting that my categories of lower animal and higher rational activity are confirmed by the pessimistic Hobbesian and the optimistic Enlightenment views of man. However, the analysis of this book is based upon careful reasoning from man's brain structure rather than on pure speculation.

Clearly, any explanation of man in society that relies on respect for and fulfilment of the human person will be in the tradition of the Enlightenment; and will therefore depend greatly upon the concept of human rationality. Durkheim wanted to stabilise society on the basis of respect for human personality and personal autonomy; but as Aron points out, 'As the emphasis is placed on social norms or on the fulfilment of individual autonomy, a conservative or rationalist-liberal interpretation of Durkheimian thought is suggested' (Aron 1968 p.106). This again demonstrates the intrusion of the two modes, the lower where sanctions and coercion are vital, as in Hobbes's theory, and the higher rational ideal which relies on personal integrity and autonomy.

There is also a bizarre tension in the work of Weber; so that his great preoccupation with rationality culminated in his pessimistic belief in the control exerted by the bureaucratic way of life as being totally compulsive for its participants. This arises from his failure to define rationality consistently; he therefore ends with different kinds of rationality which are incompatible. The incompatibility in turn arises because Weber fails to see that the rationality which defines the objectives must be common with that which is used in pursuit of those objectives. In particular, Weber appears to countenance any kind of setting of objectives whether of higher rational motivation or of lower animal type; all are grouped in his scheme under the category 'substantive rationality'. Only a carefully considered definition of rationality, taking into account both the lower and higher motivations of man, can obviate this incompatibility. The important principle involved here is that whereas rationality concerns only the higher mode, nevertheless much of man's activity and objective setting is in the lower animal mode of action.

It has been suggested that Dawe firstly, ignores the possibility that social system and social action viewpoints are simply two sides of the same coin or that secondly, maintaining social control and excercising human control over social forces are but two aspects of the general problem of securing the required situation for the development of man's capacities. However, one cannot agree entirely with either of these judgements. The first glosses over the very real source of conflict which arises from the two quite different modes of activity, the higher and the lower. The second is true as a limited statement of an ideally higher rational situation, however it again neglects the lower mode of activity which in practice prevents the required situation for developing man's capacities from ever being fully achieved.

Berger and Luckman elucidate the dual character of the world as both objective and subjective reality. It is objective in that its institutions exist externally to him; and they constrain man. Berger and Luckman show how the socialisation of a child involves problems of deviation from existing social norms, compliance and sanctions. (p.557 *Sociological Perspectives*, Thomson & Tunstall) 'The priority of institutional definitions of situations must be consistently maintained over individual attempts at redefinition. The children must be taught to behave and be kept in line, and so of course must the adults.' The conflict and need for control arises because of the difference between objective and subjective realities: and this in turn is caused by the necessity to socialise every individual commencing in infancy with animal biological instincts; and developing from these towards a socially rational state. This refinement from the lower animal towards the higher rational behaviour is a normal part of the effort to socialise and educate every individual from infancy. Even so every human being remains also an animal; therefore the higher and lower modes of activity remain interwoven in their lives.

However, as Berger and Luckman point out, 'If socialization into the institutions has been effective, outright coercive measures can be applied economically and selectively. Most of the time, conduct will occur "spontaneously" within the institutionally set channels.' This indicates the influence of self-control which is the higher rational way of reconciling the problem of order and control. However, the continuing need for some coercion emphasises also the continuing tendency for the lower animal behaviour to make inroads upon the precariously established higher rational mode of action. Thus the dichotomy between the two modes of activity remains interwoven into the fabric of man's social life.

Other sociologists have attempted to take account of ambiguity between order and control; Goffman's dramaturgical model suggests ways in which social stability can be maintained whilst allowing some degree of freedom to individual actors in negotiating to persuade each other to accept certain definitions of the situation.

Alfred Schutz and his followers are described as phenomenologists or ethnomethodologists. These see the social world held together by a dense infrastructure of tacit understandings of men concerning each other most of which have no special significance except to the individuals in question.

In the theories of Goffman and Schutz the ingredients of conflict demonstrates itself in a good deal of cheating, lying and deception performed in order to gain the advantage in negotiation. The emphasis is not on shared values maintaining social stability, but rather conventions which are treated as rules of the game, to be manipulated to the individual's own advantage.

It is, however, important to recognise that there is a difference between legitimate mutual exchange and the dishonest manipulation of the truth; the former is based on mutual regard; while the latter generates conflict. Regard for the truth lies at the basis of all rationality, co-operation and mutual regard; it is therefore also the basis for morality and the higher rational mode of activity. Thus it is the denial of this essential basis of truth that destroys the higher rational mode of action and reduces men to the status of cunning animals each fighting for his own private gain without regard for others in society.

By contrast, a social system held together by shared values was the ideal of Comte and Durkheim, and more recently of Talcott Parsons. The deficiency of these theories lies in the fact that they are very much ideals and they neglect the conflict in society that is generated by man's animal mode of activity. The theories of Comte, Durkheim and Parsons are, therefore complementary to those of Goffman and Shutz; the former tend to assume the higher rational mode while the latter assume the lower animal motivation. This clearly emphasises the importance of adopting a scientific systems model, as proposed in this analysis, which includes both of these modes of man's activity in society.

The ideas put forward by the exchange theorists are basically similar to the utilitarianism of the nineteenth century in which it is the factor of mutual usefulness that maintains the social order. However, the concept of utility is difficult to describe adequately; moreover, the concept that freedom of the individual provides scope for competition and hence the greatest good of the greatest number, is also difficult to reconcile with present day conditions; in which, the domination of large corporations and multi-nationals renders the individual insignificant.

It appears that the theory of utility was a striving towards some rational common denominator which was, at that time hard to identify. However, the essential common unifying factor is now provided by the energy analysis and the use, in this book, of scientific concepts. It is not simply man competing but also men co-operating in social groupings in which the concepts of systems, dynamic self-maintenance, morphogenesis, uncertainty, entropy and energy are all relevant.

It can be seen from the brief outline given above that the scientific analysis of this book is capable of imparting the complementarity and unification that is so essential before the great variety of different social theories may be seen to be in some way compatible.

Therefore, both the systems and the action aspects of social science can be retained and will make their own necessary contribution to the overall scientific model provided that the two modes of human action are considered and the above mentioned scientific concepts are carefully employed.

Only in this scientific way is it possible for sociology to be sufficiently comprehensive to encompass society both as a network of human actions and also the entity of society presiding over its members and moulding them with its coercive controls into its social processes.

24

SUMMARY

The Reason for the Energy Analysis
Energy was identified by Einstein as an ingredient of all material phenomena: it is a pervasive common factor which enables scientific co-ordination into one unified general theory, the profusion of theories in the specialised disciplines of the more exact sciences.

There is a similar profusion of theories in the social sciences, each motivated by some specialised viewpoint; the pervasive energy aspect of all matter also provides a medium for co-ordinating these into one general unified theory.

The relevance of energy to all these theories derives from the fact that energy is a factor in all the materials with which man must work, create and express himself; it pervades his body and his environment: therefore, he experiences the influences of C.O.E.* as a universal constraint upon all of his activities. He must consider their effect on all of his plans and objectives. Therefore the laws of C.O.E. should have, if man is to be rational, great influence in his deciding on objectives and also upon the ways in which he tries to achieve them. It follows that the principles of C.O.E. should greatly influence all man's rational relationships; and therefore also all scientific social theories that are designed to explain such relationships.

Some very brief indication of how this can be done is outlined in this book; however the detailed work will eventually fill several larger volumes. Only the overall theory is indicated here; nor is it claimed that energy is the 'sole cause'. Other principles, scientific and otherwise, are involved in social activity.

Nevertheless, if men ignore the principles of energy, that does not suspend the functioning of such natural laws. Men are not precisely and deterministically controlled by energy; but they do have to accept the consequences that arise from their own decisions which allocate their scarce available resources.

In some way, also the idea of pervasive laws is essential to the concept of rationality and hence also to science. Human social

*C.O.E. = Conservation of Energy

evolution can be visualised as conforming, ideally, to increasing rationality with the progressive expansion of understanding of the pervasive scientific laws.

A Brief Statement of the Hypothesis

Man is an animal and his bodily energies derive from common biological sources with those of his animal ancestors. However, he has the potential to refine these onto higher rational planes than those of other animals.

Therefore, two intimately interwoven models of evolution are postulated, namely, the animal biological (lower animal), and the higher rational.

This duality is the source of many ambiguities and human problems. Both modes are conditioned by pervasive energy constraints; but in the higher rational mode man becomes increasingly aware of the rational scientific principles involved. These are implemented by innovations which serve the increasing harnessing and conserving of natural energy and resources.

In the higher rational mode it is no longer the human organisms that struggle for survival; but the ideas and innovations are selected and survive only if they harness energy or conserve resources more effectively for man. The struggle for survival is, thus, transferred from the human organisms to the ideas or inventions that serve C.O.E.

There is a general increase in potential for satisfying human wants if more energy or resources are harnessed for productive purposes or if existing supplies are used more effectively. These arguments do presuppose, however, that the objectives to which the increasing energy supplies are put, are rational. This is discussed later in this summary.

Classification and Examples of Energy Harnessing Devices

It is not only machines that are necessary for harnessing energy; but also every device that helps to form the environment in which the machines must function, is a means whereby energy is harnessed. The following is a list of the various energy controlling devices with a few examples:

Machine Tools and Implements:

Axe, Knife, Harpoon, Spear, Hammer, Saw, Scraper, Needle, Potters Wheel, Spinning Wheel, Grinding Wheel, Sails,

Windmill, Agricultural Machines, Plough, Reaper, Steam, Petrol, Oil and Gas Engines, Railway, Lathe, Loom, Atomic Reactor.

Means of Communication:

Telegraph, Telephone, Radio, Television, Printing, Tape Control, Computers, Automation.

Symbolic Representation:

Speech, Writing, Mathematics, Computer Codes, Methods of Calculation, Scientific Models, Social Theories, Money.

Methods of Group Organisation:

Limited Liability, Specialisation, Division of Labour, Government, Authority, Obedience, Cartels, Multinational Enterprises, Nationalised Industries.

Intermediate Between Symbolic Representation and Methods of Group Organisation:

Rules, Regulations, Accounting Methods, Scientific Methods, Laws, Time and Motion Study, Operation Planning.

Feelings, Beliefs, Attitudes and Ideals:

Mutual regard resulting in co-operation and economies of specialisation and division of labour. (Mutual distrust and antagonism resulting in waste, conflict and destruction.)

Customs and Institutions:

Marriage, Parliament, Rituals etc.

One cannot treat the subject of energy controlling devices adequately in a short space; but it is worthy of mention that the thrust of a native's spear may concentrate his puny energy into a vital spot where it can kill the animal which he is hunting before it kills him; thus it enables him to survive.

Science in the Social Context

The principles of the more exact sciences must be read across to social science. These include: C.O.E., uncertainty, entropy and systems analysis. The scientific laws of matter are felt as constraints even though men are able to take decisions and choose courses of action.

Several hitherto ambiguous problems are elucidated by the present analysis; and these extensions of scientific knowledge will serve as empirically verified referents in the social sphere. Where there can be no deterministic predictions empirical referents in social science must be inevitably of a reflexive nature.

Rationality

A man may be rational within the limitations of his own exclusive ends; but group activity, motivated by mutual regard, is the most effective method whereby men may seek understanding and control of their environmental resources. Thus any rationality which neglects the economies and intrinsic satisfactions arising from group action represents only a small proportion of man's potential. Acting solely as individuals, men are no more than extremely cunning animals; when men are set against men in conflict, resources are wasted.

It is central to the thesis presented in this work that the lower animal, biological mode and higher rational mode of evolution are intimately interwoven into man's life. The ideal proposed by Novicow is, therefore, the best that can be aimed for; namely that conflict should be increasingly sublimated and refined onto the intellectual level.

Profligate waste of resources, pollution and warfare are irrational. These wastes arise out of antagonisms, conflict, and self-centredness. Understanding, conservation and controlling of environmental resources is promoted by co-operation, mutual regard and the economies of reciprocal service. Conservation of energy can be, therefore, identified as an aspect in the process of man's intellectual and also moral advance.

Antagonistic conflict is irrational because it is directly opposed to the spirit of enlightened and unselfish search for truth, which has been traditionally recognised as the basis of all education and science. This ideal of education promotes also, as a by product, mutual assistance and proliferation of material satisfactions together with the vital benefit of security.

Therefore, there is no need to aim for optimisation in social affairs, by machine-like precision and uniformity. It is essential to appreciate man's plasticity of behaviour; thus men may fit into complementary niches with men; and a variety of differing human cultures can interweave into an overall ecosystem existing in mutual co-operation and appreciation. This concept of humanity forming an ecosystem in harmony within itself and with its environment, is in accord with the modern conservationist view whereby waste is minimised, resources are proliferated by co-operation and planning; and pollution is prevented.

However, these are ideals for which man should aim,

nevertheless, they can never be attained to perfection because of man's dual nature in which the lower animal biological mode intermingles with the higher rational mode behaviour, providing, indeed, the resources for the higher from the raw materials of the lower.

Ideals are essential to man and the objectives which he sets may be either rational or irrational. Rational objectives enable the social system to maintain itself or to expand, though its particular form may adapt (morphogenesis). Irrational objectives will ultimately cause loss in the capacity of the social system to maintain itself; thus it will deteriorate and contract or possibly pass out of existence due to the wastages of pollution, warfare and the profligate use of resources.

Conflict is, therefore, very relevant to man's social life since it is involved in the lower instinctive processes whereby man has evolved. Conflict also exists within every individual whenever the higher rational processes strive to master and use the lower animal instinctive raw energies to pursue ideal goals; nevertheless, this fact must not be allowed to detract from the existence and indeed, the necessity for some higher ideals and more rational co-operative objectives for human strivings and ambitions. Ideals are the main distinguishing characteristic of man as a species.

Morality

When they are defined with strict logic, rationality and goodness are identical; and the moral order can be identified with the higher rational mode of human behaviour as defined in the previous chapter. However the correct priority in personal motivations is essential; that is, the right attitude of mutual regard must come first; and this will lead on to the material benefits from the economies of co-operation: any attempt to seek the material gain first is likely to end in conflict and waste of resources.

The Co-ordination of the Various Social Theories

The analysis in this book is not intended to replace previous social theories; but the scientific framework and model is shown to provide a means for co-ordinating and unifying many existing social theories into one whole.

This focuses upon the social system versus social action

controversy, which is the modern expression of the ancient 'free will' versus determinism paradox. The following have been particularly considered in the context of the energy based model with two modes of activity:

Social Action, Social Systems, Parsons, Buckley, the Fundamental Assumptions of Economic Theory, Marx, Weber, Darwin, the Utilitarians, Exchange Theorists, Psychology, Freud, Jung, Thouless, Childe, Redfield, Comte, Durkheim, Dawe, Hobbes, Berger and Luckman, Goffman, Schutz, and others.

Philosophy

In a work of social science, such as this, it is inevitable that some philosophic implications will occur. In any study in social science, certain philosophies will be implied by the scientific model used; it must be observed, however, that philosophy tends to pose problems; whereas, science has the task of elucidating them as and when it is able. For this reason it appears that one cannot prove scientific matters by reasoning from philosophic premises: nevertheless, a true philosophy can never be at variance with a valid scientific theory.

There is no shortage of philosophies; whatever theory or course of action a person might decide to take, there will usually exist a philosophy ready made to support it. Thus, Hitler found in Hegel and other German philosophers a convenient basis for his intentions.

In my view, one cannot separate philosophy from ethics and morals. Indeed, science, ethics, morals and philosophy need to be unified in one set of ideas, models and concepts. This is the ancient doctrine of 'unity in truth' or the modern concept that all study should be 'interdisciplinary'.

It is always difficult to separate an individual from the cultural influences of his social environment. My own theories are quite consciously founded upon Christian philosophies. The analysis of this book emphasises, in accordance with Christian principles, man is recognised as being able to choose his actions on rational bases. This emphasis is quite contrary to any concept of automatic determinism in human activity.

There are, I feel, some people to whom scientific method and conservation of energy simply mean absolutely automatic determinism. They appear to prefer this oversimplified, shorthand approach, despite the fact that it contains, in its compressed kernel, the seeds of all the major confusions and errors in social theory.

Such trite definitions are just not adequate or even meaningful; moreover, they are even more insidiously destructive of rational thought when such definitions exist only as mental sets or prejudices; which gain plausibility from philosophies especially selected to underpin them.

Therefore, inadequate or loose definition can lead to a great deal of discussion which is both vague and unnecessarily philosophic in its appearance without gaining any deeper insight into the problems in hand. The term rationality is an outstanding example of a concept whereby proliferation of quasi-philosophy is often caused by inadequate definition. It will be noted that the concept of rationality has, for this reason, been given a very thorough definition in this analysis.

Much of the detail in this book and the model of evolution expounded herein is scientifically verifiable and is most certainly not intended as any form of philosophy. Higher systems are seen as created from the raw materials provided by the lower systems; with rationality and variability of behaviour increasing as the hierarchy moves upwards. Man is at the apex of this hierarchy; but not just individual man; in fact, man in society conceived as an 'organism' or more scientifically as an ecosystem, stands at the pinnacle of the hierarchy.

The distinction is made between animal biological behaviour and the higher rational mode of activity. This distinction is, in my view quite fundamental to both Christianity and to social science; but although this essential distinction is emphasised in Christian teaching, it has received little systematic recognition in social science. This book attempts, in some small way, to initiate a remedy for this deficiency.

The major contribution of this work is intended to be the translation of some concepts established in the more exact sciences for use in social science, an attempt to co-ordinate the existing social theories upon the basis of these concepts, particularly upon energy; and the elucidation of some major paradoxes and ambiguities in social theory. Success in clearing up these ambiguities is intended as verification of empirical referents. If one could stand apart from man and see him objectively, then the paradox of his 'free will' versus the 'determinism' of scientific laws requires an explanation, just as surely as did the discrepancy in the orbit of Mercury in the more exact sciences. That morality is rational also requires an explanation. It is in the nature of these explanations to have a certain philosophic flavour; however, I feel that this is only an illusion of philosophy; and the whole basis and method of this work remains, as it was intended strictly scientific.

I have not intended to discuss aspects which are not of the material order. It is for philosophers and theologians to discuss what is the mysterious basis of all matter which is everywhere to be felt but nowhere to be actually seen and identified; I talk of and analyse energy but what exactly it is cannot be precisely defined.

25

EPILOGUE

One feels compelled to offer some kind of brief summary. For me this goes very much 'against the grain'. Brevity may be the soul of wit but it leaves far too much to the imagination in scientific matters!

Although some of the ideas in this book may be unfamiliar and novel, there is no part of them that is consciously intended to be contrary to Christian Catholic philosophies. It is very easy to make mistakes, particularly by oversimplifying complex matters. Some such errors will no doubt have occurred in this book; and I look forward to being able to correct them when they are pointed out. Finally I will close with a quotation from *Christ Our Eucharist* by Edward Holloway S.T.L. pp.1-2, which expresses my feelings and philosophies:

'In Christ the Word, or Living Wisdom of God in Person the entire Creation proceeded, from the energies which produced the forms of elemental matter, to the angelic nature, made in the simple image of God. Nothing came forth from the Word in chaos, but all energies poised and measured through a law of being which is a law of ascent. All things were poised in one law of control and direction, so that the laws of the sciences of matter run the one into another, in an ascending harmony of creation, the universe God made builds up like a pyramid, the broad bases upon the vast universe of matter, the apex a point in the making of Man, Man a being of spirit and of matter, Man made to the image and likeness of God, to the image and likeness specifically of the Word of God who exists from eternity, and who therefore upon his incarnation for us, chose and preferred the title of "the Son of Man".'